T0195556

THE MARINE CORROSION PROCESS AND CONTROL

Design Guides for Oil and Gas Facilities

Matthew Omotoso, Ph.D

authorHOUSE

AuthorHouse™
1663 Liberty Drive
Bloomington, IN 47403
www.authorhouse.com
Phone: 833-262-8899

Published by AuthorHouse 01/27/2022

ISBN: 978-1-7283-4769-1 (sc)
ISBN: 978-1-7283-4771-4 (hc)
ISBN: 978-1-7283-4770-7 (e)

Library of Congress Control Number: 2020903462

Print information available on the last page.

To my big uncle and early life coach, Pius Olaniyan, who is without doubt running about in heaven and showing everyone his little Matthew's book.

CONTENTS

PREFACE

The protection of marine infrastructures and pipelines against corrosion damage has become a norm within and beyond the petroleum industry, now more than ever before. This book is for those willing to enhance their knowledge in the areas of engineering and construction of vulnerable engineering installations against corrosion, as well as for students and engineers who wish to specialize in oil and gas facilities engineering and installation. It prescribes several corrosion control techniques and risk-based assessment of corroded structures. Classical examples of cathodic protection design are presented with special reference to sacrificial and impressed current principles.

Most corrosion engineers today obtained a degree in other disciplines. However, by undergoing several specialized courses of study, on-the-job training, and individual exposure to field environments, they gain knowledge and skill and become corrosion engineers. I was motivated to publish this book by the dearth of locally available printed books on corrosion written in simple and comprehensible engineering language.

The major area that has been a challenge in the current oil and gas production boom in deepwater is corrosion and materials technology, which requires innovation in the area of alternative alloys to resist corrosion and provision of higher-strength materials for subsea structures, weight-saving materials, and composites.

Marine structures are constantly exposed to seawater that hastens corrosion damage because these installations may have deteriorated to a certain degree through decades of existence in seawater. However, adequate safety of the degraded marine structures may be achieved through the reliable, risk-based assessment techniques that I will specify.

This book contains some references, specifications, and other relevant technical tables, but discussions of these provisions are not intended as substitutes for specifications and technical tables. Engineers are obliged to make direct reference to the latest edition of the particular specification relating to a given engineering design.

As a practicing engineer in the petroleum industry, I have had the opportunity to work on several upstream projects in Nigeria, France, and the United States, including deepwater project supports, subsea engineering and equipment, jacket platform design, and structural assessment. Throughout my career, I have had the privilege of working with experienced corrosion engineers, enabling me to learn about recent developments in industrial applications and research.

Matthew Omotoso, PhD

ACKNOWLEDGEMENTS

The author wishes to acknowledge the contribution of Dr. Isaac Akiije, who provided the initial editorial review of the book.

A special note of thanks to Professor David Esezobor, whose reading of the final proofs was a great help rendering some of the technical representation in a simple and distinct manner.

I really appreciate my dear wife and my wonderful children for their love and encouragement.

Finally, I thank all organizations and individuals for their support and contributions to the success of this edition.

CHAPTER 1

FUNDAMENTALS OF CORROSION

1.0 Introduction

Corrosion is degradation of materials' properties caused by interactions with their environment. The consequences of corrosion damage have become a worldwide hazard that undermines safety, economy, and conservation. Premature failure of steel structures and operating equipment as a result of damage from corrosion may cause human injury and even death. In addition to our everyday encounters with this form of degradation, corrosion wastes valuable resources, contaminates product, reduces facility efficiency, and adds to the cost of maintenance and expensive overdesign. Practically, corrosion occurs on all metallic structures that are not adequately protected from corrosion agencies. However, prompt replacement of structures that may have been damaged or weakened by undue corrosion is an important preventive measure.

In the atmosphere, the intensity of corrosion attack is influenced greatly by the amount of salt particles and moisture that collects on the metal surface. The amount of rain and its distribution during a given period of time affects the corrosion rate in marine atmospheres because frequent rain reduces the attack by rinsing off some salt residue on exposed metal surfaces. Tropical marine environments, with their high temperatures, are considered more corrosive than the arctic marine environment. In some cases, corrosion on the sheltered side of metal may be worse than that on the exposed side because dust and airborne sea-salt contamination is not washed off. Fungi and molds may deposit on metal surfaces, increasing the presence of moisture and hence the corrosion rate. Pitting corrosion takes place much faster in areas where welding operations or poor design have led to microstructural changes.

Localized corrosion of an offshore structure may provide sites for fatigue initiation that greatly enhance the growth of fatigue cracks. The elements in the splash zone are more or less continuously wet with well-aerated seawater, which greatly contributes to corrosion processes in the zone. The wind and ocean waves combine to create violent seawater conditions; impinging water also complements the corrosion rate for marine structures in the splash zone.

Of all the marine zones for steel materials, the splash zone is most subject to aggressive corrosion. The presence of air bubbles in the seawater critically removes the protective films or dislodging coatings on the marine structures. Consequently, paint films normally deteriorate more rapidly in the splash zone than in other zones.

Corrosion damage can be classified according to the geometry and the reaction that lead to its formation. The most important classes of corrosion include uniform corrosion, pitting corrosion, crevice corrosion, intergranular corrosion, stress corrosion, and corrosion fatigue. One type of corrosion may militate against another. Along similar principles, all forms of corrosion are divided into three groups, depending on the method of identification. The types of corrosion that fall within group one is readily identifiable by ordinary visual examination. The corrosion that requires a means of examination in addition to visual inspection belongs to group two. Group three corrosion requires verification by techniques of microscopy such as optical or scanning electron.

The consequences of any type of corrosion are many and diverse. Its global effects on the safe, reliable, and efficient operation of equipment and structures are far graver than the simple loss of a mass of metal. Different forms of corrosion failure may need repair and expensive replacements, even though the amount of metal damage is relatively minute. For that reason, the major harmful effects of corrosion in day-to-day activities on engineering facilities include the following.

- Decrease of metal thickness and cracking, leading to loss of mechanical strength that may result to structural failure or collapse
- Structural failure and the rupture of products pipeline that may lead to hazards or injuries
- Loss of time in availability of profile-making industrial equipment
- Reduction of metal value and unappealing appearance due to deterioration
- Contamination of fluids in vessels and pipelines by corrosion products
- Pitting corrosion occurring in vessels and pipelines that may allow their contents to escape and possibly contaminate the surrounding area
- Loss of technical properties of a metallic component, including frictional and bearing properties, product flow rate in a pipeline, electrical conductivity of contacts, and surface reflectivity
- Mechanical damage to pumps and valves, and eventual blockage of pipes by solid corrosion products
- Increase in facilities cost and equipment complexity that requires designs to withstand a certain amount of corrosion losses

Addressing the above concerns and studies has led to the development of new metal alloys and many nonmetallic construction materials, including a wide range of thermoplastic materials as well as several varieties of coatings and linings.

Prior to the material selection for a particular application, it must be determined that the material has physical, mechanical, and corrosion-resistant properties. Cost implications must also be considered in construction material selection—there may be many alloy metals available to meet design criteria, but the most economical must be selected. Consequently, various coating and lining materials have been developed for application to less expensive construction materials in order to meet the required corrosion resistance.

Given all the factors mentioned earlier, it is essential that the potential problem of corrosion be given adequate attention during the early design stage of oil and gas facilities projects. It is also necessary to continuously monitor the integrity of structures and equipment throughout the life span of the facilities to prevent corrosion failure.

Managing the potential problem of corrosion in every project requires a thorough understanding of the following aspects of corrosion phenomenon, which shall be covered in this book.

i.   Corrosion theory and mechanism
ii. Forms of corrosion
iii. Corrosion model
iv. Corrosion risk assessment
v.   Corrosion monitoring techniques
vi. Corrosion control methods
vii. Cathodic protection systems
viii. Cathodic protection design examples
ix. Marine structure corrosion damage
x.   Corrosion safety and economics

1.1 Corrosion Theory and Mechanism

The term *corrosion* describes an electrochemical reaction that takes place during the reaction of material with its surroundings causing the metal damage. Corrosion is the primary means by which metals deteriorate when in contact with the water, moisture in the air, acids, bases, and salts that are normally present in an environment. For example, iron and steel have a natural tendency to combine with other chemical elements to return to their lowest energy states. In order to return to these lower energy states, iron and steel frequently combine with oxygen and water, both of which are present in most natural environments, thereby forming hydrated iron oxides, popularly known as rust, which are similar in chemical composition to the original iron ore. The life cycle of a steel product is illustrated in figure 1.1.

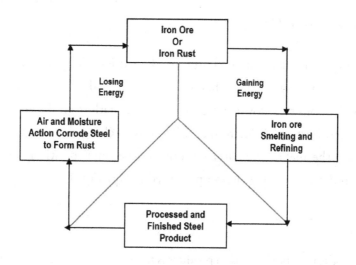

Figure 1.1. Steel product corrosion life cycle

Corrosion in aqueous conditions is the most common of all processes that are electrochemical in nature. Water into which have been dissolved various salts and gases, seawater, or processed streams from industries is rendered capable to some extent of conducting and acting as an electrolyte. The chemical nature of this electrolyte may be acidic, alkaline, or neutral.

An electrolyte is a solution that must be present before corrosion occurs. The ions within an electrolyte enable it to conduct electricity.

When dissimilar metals that differ in electrical potential come in contact with an electrolyte, a continuous electrical path or corrosion cell is built between the two metals. The common types of electrolyte solutions in oil and gas facilities are salt, acid, and hydroxide. These chemical solutions are highly conductive and have a tendency to accelerate corrosion cell action. Iron metals linked in an electrolyte are subjected to the following typical reactions. The cathode in a corrosion cell is the area of metal immersed in electrolyte that contains surplus negative electrons with a negative charge. The positive hydrogen ions absorb the negative charges from the cathode and become molecular hydrogen, as shown in equation 1.1.

$$2H^+ + 2e^- \longrightarrow H_2 \tag{1.1}$$

The giving away of negative charge by the cathode lowers its electrical potential compared to the other part of the metal. The cathode then collects electrons that flow through the metallic path from the anode that now creates shortage of electrons at the positively charged point in the cell. Therefore, the metal gives up electrons and becomes positively charged ions, as indicated in the reaction presented in equation 1.2.

$$Fe - 2e^- \longrightarrow Fe^{++} \tag{1.2}$$

At the anode, the above reaction product is not stable, and therefore a second reaction will quickly take place in which iron ions react with hydroxyl ions that are present in the electrolyte, as symbolized in equation 1.3.

$$Fe^{++} + 2(OH)^- \longrightarrow Fe\,(OH)_2 \tag{1.3}$$

The hydroxide product is insoluble and hence separates from the electrolyte. Anodic reactions in metallic corrosion are somehow simple because the reactions are always such that the metal is oxidized to a higher valence state.

During the corrosion process, it should be noted that the dissolution process of the metal is taking place through anodic reaction, and the electrons liberated by anodic reaction are consumed in the cathodic process. The metal that is going under the corrosion process does not accumulate any charge because the two partial reactions of oxidation and reduction proceed simultaneously and at the same rate to maintain the electro-neutrality. Several vital models of corrosion protections were developed based on this electrochemical representation. Retarding the cathodic process can retard metal dissolution by supplying electrons to the corroding metal from any external source, which forms the basis of cathodic protection using the impress current system.

1.2 Atmospheric Corrosion Mechanism

Atmospheric corrosion can be simply defined as the corrosion of materials exposed to air and its pollutants. The rate of atmospheric corrosive attack is dependent on the properties of the thin-film surface electrolyte. Most metal corrosion occurs via electrochemical reactions at the interface between the metal and an electrolyte solution. A thin film of moisture on a metal surface forms the electrolyte for atmospheric corrosion. This type of corrosion has been identified to be responsible for more failure in term of cost and material wastage than any other corrosion mechanism. The atmospheric corrosion can be grouped into either damp or wet. The damp moisture films on materials are created at a certain humidity level, and wet films are caused by dew, ocean spray, rainwater, and more.

Atmospheric corrosion is an electrochemical process, requiring the presence of an electrolyte. Thin, invisible film electrolytes tend to form on metallic surfaces under atmospheric corrosion conditions when a certain critical humidity level is reached. In the case of iron, this level is about 60 percent in unpolluted atmospheres. The critical humidity level depends on the corroding material, the hygroscopic nature of corrosion products and surface deposits, and the presence of atmospheric pollutants.

In the presence of thin film electrolytes, atmospheric corrosion proceeds by balancing anodic and cathodic reactions. The anodic oxidation reaction involves the dissolution of the metal in the electrolyte, whereas the cathodic reaction is considered to be the oxygen reduction reaction. The anode and cathodic reactions involving iron and oxygen are presented in equations 1.4 and 1.5, respectively.

$$2M \longrightarrow 2M^{2+} + 4e^- \qquad\qquad\qquad (1.4)$$

$$O_2 + 2H_2O + 4e^- \longrightarrow 4OH^+ \qquad\qquad (1.5)$$

Atmospheric iron corrosion is illustrated schematically in figure 1.2.

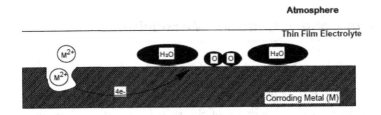

Figure 1.2 Atmospheric corrosion

The corrosive pollutant concentration can be higher in the thin electrolyte films under the condition of alternate wetting and drying as it applied to the splash zone in an offshore environment. It should be noted that oxygen from the atmosphere is readily supplied to the electrolyte under thin film corrosion conditions.

Electrochemical corrosion consists of two half-cell reactions, the oxidation reaction at the anode, and the reduction reaction at the cathode. Electrochemical corrosion occurs in the presence of four fundamental components: an anode, a cathode, and electrolyte and electrical collection between anode and cathode to allow flow of electrons. In the absence of these mentioned components, electrochemical corrosion will be categorically stopped, which makes these elements essential for corrosion control.

Corrosion thus occurs at a rate determined by equilibrium between opposing electrochemical reactions. The first is the anodic reaction, in which a metal is oxidized, releasing electrons into the metal. The other is the cathodic reaction, in which a solution species is reduced, removing electrons from the metal. The two reactions usually take place on a metal or on two dissimilar metals that are electrically connected.

Metallic materials consist of one or several metallic phases, however this depends on their composition. In the metallic state, atoms donate some of their outside electrons to the electron gas that is responsible for the good electrical conductivity of the metal. It is known that pure metal does not react electrochemically as a single component, and therefore an assumed approximate iron state during some phases of corrosion process is presented in equation 1.6, where iron ions and electrons are present in the metal.

$$Fe \longleftrightarrow Fe^{2+} + 2e^- \hspace{3cm} (1.6)$$

The two elements can react with electrolytes. The dissolution of the metal by the passage of Fe^{2+} results in a positive current I_A and a metal loss of Δm. This reaction is equivalent to anodic reaction because electrons are transferred from the metal. Also, the passage of electrons leads to a negative current I_C without metal removal but deposited of iron, which is a cathodic reaction. Thus, during electrolysis process, cathodic metal deposition is making use in electroplating, whereas this is a reverse process during electrolytic corrosion, when metal is loss in anode. Therefore, Faraday's law of electrolysis is applicable to the both processes, as stated in equation (1.7).

$$\Delta m = \frac{MQ}{zF} \hspace{3cm} (1.7)$$

In equation 1.7, Δm is the mass dissolved metal, M is the atomic weight, Q is the transferred electric charge, z is the valence of the metal irons, and F is Faraday's constant. The subsequent equations can be derived using the specific gravity ρ_s:

$$J_A = \frac{Q}{St} = \frac{\Delta m}{St}\frac{zF}{M} = \frac{\Delta s}{t}\frac{zF\rho_s}{M} \qquad (1.8)$$

Here, J_A is current density of the anodic partial reaction for the passage of metal ions, S is the surface area of the electrode, t is time, and Δs is the thickness of material removed.

The rate of corrosion tends to differ considerably among different locations because it is applicable to onshore and offshore environment. Different local material removal rates are often due to difference in nonuniformity of the surface films, where both thermodynamic and kinetic effects have influences. Therefore, uniform thickness loss of metal happens mostly on active and in a single-phase metal. Selective corrosion is mostly common on multiphase alloys.

1.3 Dissimilar Electrode Cell

The formation of the dissimilar electrode cell occurs when two dissimilar metals are in contact. In engineering practice, a copper pipe connected to a steel pipe or steel hull of a ship is an example of the corrosion cell. A standard corrosion cell consists of metals that act as anode or cathode, electrolyte, and conductor between the electrodes to allow current to flow. An example of an electrolytic cell in which zinc behaves as anode and copper act as cathode is illustrated in figure 1.3.

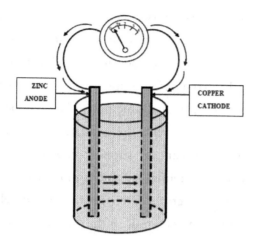

Figure 1.3 Zinc-copper cells

Galvanic cell occurs when two or more dissimilar metals are brought into electrical contact under electrolytes. A galvanic couple is formed when one of the metals in the couple becomes the anode and corrodes faster than it would by itself, whereas the second metal becomes the cathode and corrodes slower than it would alone. However, either metal in the couple may or may not have corroded itself in electrolytes. The driving force of a galvanic cell is equal to the potential difference between the two metals in the cell.

Corrosion is the degeneration of metal through an unpremeditated chemical or electrochemical action, which starts at its surface. All metals exhibit a tendency to be oxidized, however some are more easily oxidized than others. A tabulation of the relative strength of this tendency, called the galvanic series can be found in the other available literature books. Knowledge of a metal's location in the series is an important part of information to make decisions about its potential usefulness for any applications. A metal will corrode if it is together with any other metal listed higher above it in the electromotive series table. The greater the separation between listed metals, the stronger the corrosion will be. Mathematically, if the galvanic potential between two metals is positive, then, the "second" metal will corrode. A negative value means that the "first" metal will corrode, and the greater the value, the stronger the corrosion.

For example, when comparing copper to a more active metal such as zinc in a copper-zinc cell, the galvanic potential of copper and zinc are (+0.34v) and (-0.76v), respectively. The standard for the reaction is then (+0.34v) – (-0.76v) = 1.10v. Zinc is a stronger reducing metal than copper metal, and the standard potential for zinc is more negative than that of copper. Zinc metal loses electrons to copper ions and develops a positive charge. Zinc will galvanically corrode in the presence of copper.

1.4 Concentration Cell

Electrolytic cells using the same metals for electrodes and cause current to flow are popularly referred to as concentration cells. The cell is in three forms: salt concentration, differential temperature, and differential oxygen content or oxidation concentration. Concentration cell is formed when electrodes are each immersed in solution with different concentration of the same salt of a metal, as illustrated in figure 1.4.

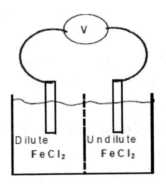

Figure 1.4 Salt concentrated cell

A concentration cell acts to dilute the more concentrated solution and concentrate the more dilute solution, thereby creating a voltage as the cell reaches equilibrium. This is achieved by transferring the electrons from the cell with the lower concentration solution to the cell with the highest concentration solution. However, the electromotive force produced by the cell will depend on the metal and the concentrations of the solutions. The differential temperature cell is not so common, like the salt concentration cell and oxygen content cell, as previously mentioned. The differential temperature cell is when the only current generation is caused by the difference in temperature around each of the electrodes, as shown in figure 1.5.

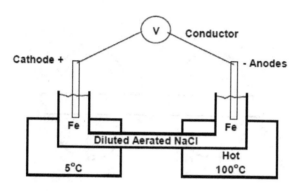

Figure 1.5 Differential temperature cell

The oxidation concentration cell is likely to be the most common concentration cell affecting engineering structures due to dissolved oxygen. A water solution in contact with the metal surface will normally contain dissolved oxygen. An oxygen cell may develop at any point where the oxygen in the air is not allowed to uniformly diffuse into the solution. However, it is promoted mainly where the oxygen concentration is the least.

In the presence of water, a high concentration of metal ions will exist under faying surfaces, and a low concentration of metal ions will exist adjacent to the crevice created by the faying surfaces. An electrical potential will exist between the two points. The area of the metal in contact with the low concentration of metal ions is cathodic and protected, and the area of metal in contact with the high metal ion concentration will be anodic and corroded. This condition can be eliminated by sealing the faying surfaces in a manner to exclude moisture. Proper protective coating application with inorganic zinc primers is also effective in reducing metal surface corrosion. The mechanisms of oxidation concentration cell corrosion in the pipeline are created by the formation of anodic and cathodic sites on the surface of the pipeline due to the existence of differences in electrical potential on the surface; see figures 1.6.

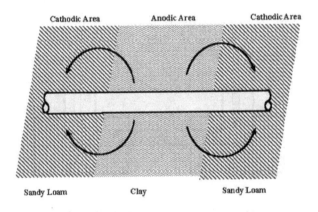

Figure 1.6 Oxygen deficiencies in the clay area

The potential differences are caused due to an oxidation concentration cell, in which the oxygen concentration in the electrolyte varies from place to place on the pipeline surfaces. A worthy example of this phenomenon is when an underground pipeline passes through clay and gravel with high oxygen concentration in the gravel region and almost no oxygen in the impermeable clay. The part of the pipe in contact with the clay becomes anodic and suffers corrosion damage.

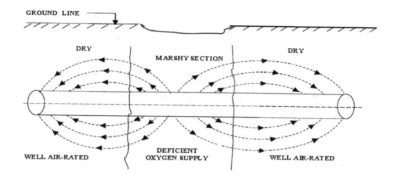

Figure 1.7 Oxygen deficiencies in the marshy area

In the same principle, a buried pipeline that passed through a mucky soil contains less oxygen than the porous and dry soil (such as gravel and loam), as presented in figure 1.7. In this case, it should be noted that the flow of current through the electrolyte was from anode to cathode, and the anode loses material in the form of ions. Therefore, the pipe section in a mucky area suffers corrosion damages while the pipe section in the well-aerated soil is not corroded.

1.5 Stray Current Electrolysis

Stray current electrolysis, or stray current corrosion, refers to corrosion caused by current leakage from an electrical system or application of cathodic protection that flows through paths other than the intended circuit. This manner of current is known as stray current due to its inherently unintentional nature. Stray current corrosion is different from natural corrosion because it is caused by an externally induced electrical current and is independent of oxygen concentration and other environmental factors. The process of stray current corrosion is electrolysis in nature. The extent of damage is directly proportional to the magnitude of stray current passing through the system. The source of electric current that provides the driving force for a stray current cell is from an outside source, as illustrated in figure 1.8.

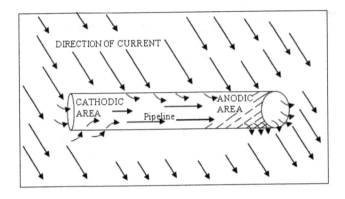

Figure 1.8 Stray current in buried pipeline

The metallic structure that is embedded in electrolyte tends to collect current at a point nearest to the current source and releases it back to the electrolyte at points far from the source, as illustrated in figure 1.8. The area of the metal structure, which gathers electrical current, becomes the cathode; the metal structure area that released the current is anodic and suffers corrosion damage.

The greatest sources of stray current in the township area are electric railway and streetcars. In oil and gas facilities, paradoxically, the common sources of stray currents are cathodic protection units, electric welding machine, and grounded DC electric sources. The use of cathodic protection system can interfere with other structures located within the device vicinity. An example of this phenomenon is the crossing of two pipelines in which one is cathodic protected and the other is not. Another example is a ship docked at a cathodically protected jetty or a pipeline attached to a cathodically protected tank. In all the cases mentioned, the cathodic protection systems provide the stray current that accelerates the corrosion damage of the foreign structures.

Stray current can be detected on underground structures by the measurement of potential using a volt difference meter of high internal resistance (minimum 100k Ω/v), as illustrated in figure 1.9. One of

MATTHEW OMOTOSO, PH.D

the outputs of the meter is connected to the structure with cable; the second input of the meter is connected to the reference electrode, which is placed on the surface of the ground as close as possible to the structure. However, modern microprocessor digital voltmeters and digital recorders are commonly used.

The potential of a structure measured versus the reference electrode has a constant value when there is no stray current around the structure, which is known as stationary potential. The potential of a structure will suddenly change in the areas of incidence of stray current due to induced polarization. The more the potential variety, the greater intensity of the stray currents in the area of measurement and the higher the probability of the current harmful interaction. Corrosive stray current interaction with underground structures that contain aggressive material, such as petroleum product in the pipelines and tanks, can lead to leakage and pollution of the environment.

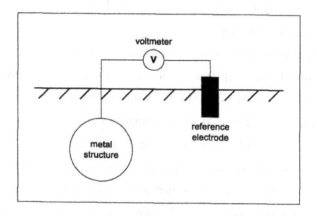

Figure1.9 Potential measurement of structures

However, the effective working corrosion protection installations will minimize the hazard. The correct choice of protection methods requires knowledge of the corrosion hazard magnitude. There are three basic approaches to the stray current dilemma prevention, as categorized below, which shall be discussed in detail in the subsequent chapter.

- Using bonding to afford a metallic return of current collected by a foreign line
- Carrying out a design that aims at minimizing exposure
- Putting in place auxiliary drainage of the collected current

1.6 Corrosion of Concrete and Reinforcement

Concrete is a composite material made of aggregates and the reaction product of the cement and the mixing water. The composition of cement paste and the environmental condition determines the durability and the long-term performance of concrete structures. Concrete is normally reinforced with steel bars to increase the structural performance. The concrete provides protection to the embedded steel to improve its ability to withstand corrosion degradation.

During the process of reinforcing steel corrosion, at the later stage, rust stains may begin to appear on the concrete surface. As a consequence, the point along the concrete surface with the highest concentration of the corrosion product may develop internal stress due to growing solid corrosion products with cracks, voids, and displacement.

Marine concrete structures are highly susceptible to corrosion due to the presence of chlorides in the seawater. Chloride-induced steel corrosion in concrete is ions dissolved in the pore solution, and critical content depends on the concrete pH and oxygen content. Carbon dioxide can also initiate reinforcement corrosion if the concrete cover is in contact with steel that is carbonated and wet, just like corroded steel without concrete cover, as illustrated in figure 1.10.

Figure 1.10 Corroded steel in concrete

Typically, steel in cement mortar is in a passive state and may act as a foreign cathodic object whose intensity depends on the aeration. However, the steel passivity may not be difficult to sustain due to the reaction of mortar with carbon dioxide, forming carbonates that significantly lower the mortar pH. Waters with an excess of free carbon dioxide attack cement mortar, and there are two steps in the mechanism to carbonization process, as illustrated by equations 1.9 and 1.10, respectively. The reaction in equation 1.14 makes the mortar lining soften and vanish. The carbonization causes the mortar to be hardened, and the steel passes from passive to active condition and starts corrode.

$$CaO + CO_2 \longrightarrow CaCO_3 \tag{1.9}$$
$$CaCO_3 + CO2 + H_2O \longrightarrow Ca^{2+} + 2HCO_3 \tag{1.10}$$

For carbonation-induced corrosion, the service life (t_l) is expressed as the sum of the initiation (t_i) and propagation (t_p) periods up to the threshold, at which deterioration becomes unacceptable and the initiation time (t_i) is a function of concrete properties.

$$t_1 = t_i + t_p \qquad\qquad (1.11)$$

When corrosion of a steel bar occurs, owing to chloride ingress in concrete, the service life may be assumed to be the same as initiation time. The period of propagation might be very short and may not be taken into account due to ambiguity of localized corrosion consequences. The corrosion initiation time may be estimated as a function of chloride transport properties of concrete, the surface chloride content, and the concrete cover. The advent of the critically chloride content on the surface of the steel at depth x and at time t can be calculated using Fick's second law of diffusion, which is explicitly discourse in Chapter 4 of this book edition.

The course of steel corrosion in concrete can sometimes be a result of the electrochemical process. The electrochemical potentials that form the corrosion cells may be generated in two ways: (1) When two dissimilar metals are embedded in concrete, such as steel rebar and aluminum conduit pipes, or when significant variations exist in surface characteristics of the steel. (2) In the vicinity of reinforcing steel, concentration cells may be formed due to differences in the concentration of dissolved ions, such as alkaloids and chlorides. As a result, one of the two metals (or some parts of the metal when only one type of metal is present) becomes anodic, and the other becomes cathodic. The fundamental chemical changes occurring at the anodic and cathodic areas are as follows.

Anode: $Fe \longrightarrow 2e- + Fe^{2}+$ $\qquad\qquad (1.12)$

Cathode: $(1/2)O_2 + H_2O + 2e- \longrightarrow 2(OH)-$ $\qquad\qquad (1.13)$

Anodic reaction (involving ionization of metallic iron) will not progress far unless the electron flow to the cathode is maintained by the consumption of electrons. For the cathode process, the presence of both air and water at the surface of the cathode is necessary. A typical corroded steel in concrete is shown figure 1.11.

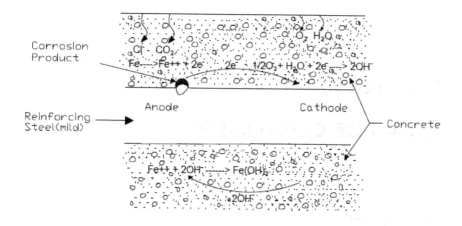

Figure 1.11 Electrochemical process of corrosion of steel in concrete

Stray current effects with reinforced concrete are not likely from the usual causes, but this is possible with roadways and bridges over which railways pass. However, in the case of stray current escape, the appropriate protective measures should be put in place to include all the DC railways and high DC power available in the concrete structures' vicinity.

Concrete contains large amounts of calcium and a small percentage of sodium and potassium ions that maintain a high alkalinity level. Steel exposed to concrete's high alkalinity forms a thin, passive film on its surface that resists corrosion. The passive film formed of steel rebars in concrete is believed to be a few nanometers thick and is primarily composed of iron oxides and hydroxides. However, little is known about the accurate chemical composition and the structure of the passive film itself. It's the failure of that film that leads to corrosion and deterioration in bridges, buildings, platforms, tunnels, and concrete pipes.

CHAPTER 2

FORMS OF CORROSION

2.0 Introduction

Designing of an effective corrosion control into components and systems requires a proper consideration to the forms of corrosion that may occur during the life span of the component. However, one form of corrosion can lead into a series of deterioration reactions where other forms of corrosion attacks may become the controlling factors.

The different forms of corrosion can be categorized according to the nature of the corrodent, the mechanism of corrosion, and appearance of the corroded metal. Corrosion can be classified as "wet" or "dry." Liquid or moisture is required for the former, whereas dry corrosion usually involves reaction with high-temperature gases. The mechanism of corrosion can be either electrochemical or direct chemical reactions. Corrosion appearance can also be uniform because metal corrodes at the same rate over the entire surface, or localized corrosion as in the case of small areas of metal affected by corrosion.

Corrosion is a major problem for steel structures and is more predominant in the marine environment. Nevertheless, when corrosion protection devices are properly applied, there should be no major problem with steel structural safety. Forms of corrosion are influenced by physical, chemical, and biological factors. Guedes Soares. C and Garbatov. Y (1999) have observed that the corrosion wastage increases nonlinearly in a period of two to five years of exposure, and afterward it becomes relatively constant. The general monograph of steel corrosion provided the fundamental basis for an extensive overview on practical aspects of corrosion forms.

Consideration of geometrical characteristics of corrosion is the first approach to the corrosion complex problem that facilitates the stochastic description of corrosion forms. Corrosion phenomena can be distinguished by geometrical characteristics without regard to the driving mechanisms. In simplified form, corrosion geometry is described as either uniform or localized, and most corrosion deterioration problems encountered in real life are a combination of these two forms. Corrosion forms (either uniform or localized) are based on their appearance on the surface of corroded metal. Uniform corrosion is macroscopic in nature and can be verified by visual examination. Localized corrosion is categorized into macroscopic and microscopic, as revealed in figure 2.1.

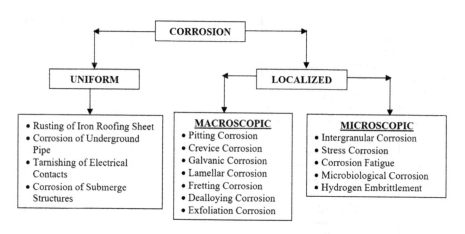

Figure 2.1 Classification of corrosion

Macroscopic forms of corrosion affect greater areas of corroded metal and are generally observable with naked eye or can be viewed with the aid of a low-power magnifying device. In the case of Microscopic Localized

Corrosion, the amount of metal dissolved is minute, but considerable damage can be done before the corrosion becomes visible to the naked eye. The various forms of corrosion that will be discussed in this book based on their mode of examination are listed in Table 2.1

Table 2.1 Corrosion forms and verification methods

S/N	Corrosion Forms	Group	Verification Method
1	Uniform Corrosion	I	Visual Examination
2	Pitting Corrosion		
3	Galvanic Corrosion		
4	Crevice Corrosion		
5	Lamellar Corrosion		
6	Intergranular Corrosion		
7	Stray Current	II	Visual and Supplementary Examination
8	Erosion Corrosion		
9	Erosion Corrosion		
10	Cavitation Corrosion		
11	Fretting Corrosion		
12	Dealloying Corrosion		
13	Stress Corrosion	III	Verified by Microscopy
14	Corrosion Fatigue		
15	Hydrogen Embrittlement		
16	Microbiologically Influence Corrosion		

2.1 Uniform Corrosion

Uniform corrosion is characterized by a corrosive attack taking place evenly over the entire surface area at a steady and often predictable rate. The corrosion process proceeds as a result of corrosion products' solubility at the rate determined by the electrode potential of the metal in the particular environment and the kinetics of the system.

Uniform corrosion occurs without appreciable localization of attack, and it is reasonably easy to measure and predict, making catastrophic failures relatively uncommon. However, uniform corrosion permitted to continue for a long time on a component surface may become rough and degenerate to more serious forms of corrosion. A typical uniform corrosion of an offshore corroded conductor guides is shown in figure 2.2.

Figure 2.2 Uniformly corroded conductor guides

Uniform corrosion can be reduced or prevented by using any of the following techniques, or a combination of them.

- Slowing down or blocking the movement of electrons on the component surfaces

- Coating the component surfaces with a nonconducting medium using paint, oil, lacquer, etc.
- Reducing the conductivity of the solution in contact with the metal surfaces and keep the component dry
- Applying cathodic protection
- Reducing or stopping oxygen from reaching the metal component surfaces
- Preventing the metal from giving up electrons by using a more corrosion-resistant metal higher in the electrochemical series
- Selecting a metal that forms an oxide that forms a protective film on metal surfaces and stops the reaction.

2.2 Localized Corrosion

Localised corrosion is based on the same basic principle of the corrosion cell as uniform corrosion, although the driving potential may be provided by different mechanisms. Localized corrosion is characterized by the fact that the anode and cathode are positioned at different locations in the structure. Furthermore, the cathode may be orders of magnitude larger than the anode, thus greatly increasing the rate of corrosion at the anode. Localized corrosion phenomena can be differentiated by their geometrical characteristics, as discussed in the following headings.

2.3 Pitting Corrosion

Pitting corrosion is a localized form of corrosion that is limited to a specific and relatively small area of a component. This form of corrosion occurs in materials that have a protective film, such as when a coating breaks down. The exposed metal area gives up electrons easily, and the reaction quickly initiates in tiny pits, with localized chemistry supporting the corrosion attack. Pitting corrosion takes the form of a deep, narrow, corrosive attack front that often causes rapid penetration of component wall thickness.

Pitting is considered to be more dangerous than uniform corrosion damage. Pitting corrosion is very difficult to predict, design against, and detect because corrosion products often cover the pits. Unpredictably, a small and narrow pit with minimal overall metal loss may lead to the failure of an entire engineering system. Furthermore, corrosion pits may also be harmful to engineering structures by acting as a stress developer, leading to fatigue and stress corrosion cracking that can be initiated at the base of corrosion pit.

2.4 Crevice Corrosion

Crevice corrosion is the localized form of corrosion and typically occurs in a constricted area of structures where free access to the surrounding environment is restricted. An oxygen concentration cell is created when oxygen cannot penetrate a crevice in the sharp edge of the structure. As a result of the differences in oxygen concentration on the steel surface, a driving potential force is produced, and corrosion commences in the area of less oxygen. This form of corrosion is also coupled with a stagnant solution on the microenvironmental level, such as those formed under gaskets, washers, and disbanded coatings. Crevice corrosion is created due to the changes in local chemistry within the crevice, like depletion of oxygen, a shift to acid conditions, and a buildup of aggressive ion species, chlorides, and breakdown of the passive film in the crevice area.

2.5 Galvanic Corrosion

Galvanic corrosion refers to corrosion damage induced when two or more dissimilar metals are brought into electrical contact with an electrolyte. As soon as a galvanic couple forms, one of the metals in the couple becomes the anode and corrodes faster than it would by itself, whereas the other becomes the cathode and corrodes slower than it would perform alone.

The galvanic series determines the nobility of metals and semimetals. Whenever two submerged metal in an electrolyte are in contact, the less noble metal will experience galvanic corrosion. The rate of corrosion is determined by the electrolyte and the difference in nobility. The difference can be measured as a difference in voltage potential. The less noble metal is the one with a lower electrode potential or the one that is more negative than the nobler one, and it will function as the anode.

In contrast to impressed current anodes, the possibility of using galvanic anodes is limited due to their electrochemical properties. The rest potential of the anode material must be sufficiently more negative than the protection potential of the object to be protected so that an adequate driving voltage can be maintained. Therefore, aluminum and zinc in seawater with reduction potential of (-0.79) and (-1.30), respectively, are suitable anodes for carbon steel or cast iron with a potential of (-0.60).

The extent of accelerated corrosion resulting from galvanic coupling is influenced by the nature of the environment, the potential difference between the metals, the polarization behavior of the metals, and the geometric relationship of the component metals. The driving force for corrosion is a potential difference between the metals. The potential of each metal is affected by environmental factors, and no one value can be given for a particular metal. A galvanic series of metal and alloys is useful for predicting galvanic relationships measured in a specific electrolyte. Therefore, measurement of a galvanic series is required in an environment of interest to determine which of the metal in the galvanic couple is more active.

Some methods for preventing galvanic corrosion is by chosen metal combinations in which the constituents are as close as possible in the corresponding galvanic series. Also, where possible, make use of a seal, insulator, and coating to avoid direct contact between the two different metals.

2.6 Lamellar Corrosion

Lamellar corrosion (LC) is a localized and subsurface corrosion in zones often parallel to the surface that results in leaving thin layers of uncorroded metal and giving rise to a layered appearance similar to the pages of a book. The corrosion commences along the grain boundaries and proceeds laterally from the point of commencement along planes parallel to the surface to form corrosion products that force metal away from the body of the material. Lamellar corrosion in ferrous alloys is associated with the excessive internal growth of the metal oxide, which sometimes exceeds that of steel by several times. This corrosion can occur in the austenitic stainless steels as a result of the thermal expansion difference between the oxide and the metal. Preventive measures against lamellar corrosion include the use of stabilized materials and heat treatment control to avoid temperature range. An example of lamellar corrosion damage of a carbon steel component is shown in figure 2.3.

Figure 2.3 Carbon steel lamellar corrosion

2.7 Erosion Corrosion

Erosion corrosion is a class of rusting that speeds up the rate of attack in a metal due to the relative motion of a corrosive fluid on a metal surface. Erosion corrosion can occur in flowing liquid or gas containing no abrasive particles. In this form of attack, the fluid is sufficient to remove weakly adhered corrosion surface, thereby reduce their polarizing or inhibiting effect; this results in an acceleration of the corrosion process of the component. The increased turbulence caused by pitting of the internal surfaces of a tube can result in rapidly increasing erosion rates and eventually a leak. A combination of erosion and corrosion can lead to extremely high pitting rates.

The fluid velocity in erosion corrosion of liquid may be an important factor for determination of material removal rate. The speed at which erosion corrosion begins in a given metal is known as breakaway velocity, and it varies from one material to another. The resistance of the protective films to the removal of flowing liquid plays a significant role in determining the breakaway velocity of a given metal in a given environment. Other factors contributing to erosion corrosion rate are particle velocity, hardness, angularity, and the temperature of the environment. Erosion by particle impact is the process by which hard particles induce corrosion during movement. In the same way, liquid particles can also induce erosion, which is generally referred to as water drop impingement. Water drop impingement is the mechanical erosion in which the eroding medium is of water drops with high velocity. The common example of such a phenomenon is raindrop erosion, which is a practical abrasion associated with high-speed jets and helicopter rotor blades.

Erosion corrosion can be prevented by an increase in the thickness of vulnerable areas. Abrasive particles in fluids should be removed by filtration. Inhibitors are additional measures that can be taken into consideration. Cathodic protection and the application of protective coatings may also reduce the rate of the erosion corrosion attack.

2.8 Cavitation Erosion

Cavitation corrosion is a mechanical damage that occurs when a fluid's operational pressure drops below its vapor pressure, causing gas pockets and bubbles to form and collapse. Cavitation is frequently induced by a reduction in the cross-sectional area of a flow passage, causing a decrease in fluid pressure and vapor gas bubble formation.

The value of the pressure differential will suggest the severity of the cave-in and the damage to the metal surface. This phenomenon can occur and even cause an explosion, produce steam at the suction of a pump, and create an airlock that may prevent incoming fluid. The locations where this is most likely to occur include the suction pump, the discharge valve, and pipe elbows. This form of corrosion may destroy valve seats and contribute to erosion corrosion, as found in the pipe elbows.

The frequently recommended prevention methods for cavitation erosion prevention includes hydrodynamic pressure gradient reduction, avoiding pressure drops below the vapor pressure of the liquid, and preventing air ingress. The use of resilient coatings and cathodic protection can also be considered as supplementary control methods.

2.9 Fretting Corrosion

Fretting corrosion is as a result of abrasive and little relative motions between the contacting surfaces. The plastic deformation at each impact extrudes metal around the particles, and the extrusions are broken off by impacts. This damage usually found in machinery, bolted assemblies, and ball and roller bearings. Contact surfaces exposed to vibration during transportation are exposed to the risk of fretting corrosion. This damage is induced under load and in the presence of repeated relative surface motion, as illustrated in figure 2.4. The most common type of fretting phenomenon is caused by vibration where the protective film on the metal surfaces is removed by the rubbing action, exposing the bare metal to the corrosive conditions.

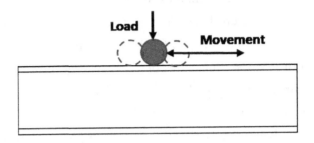

Figure 2.4 Fretting corrosion

2.10 Intergranular Corrosion

The microstructure of metals and alloys is made up of grains, which are separated by grain boundaries. Intergranular corrosion is a localized attack along the grain boundaries and sometimes is immediately adjacent to grain boundaries of a metal, whereas the bulk of the grains may remain largely unaffected. Such preferential attack along grain boundaries may dislodge the grains, leading to a decrease in efficient cross section of the metal and causing product pipe leakage or structural member failure.

Intergranular corrosion is usually associated with chemical segregation effects and specific phases precipitated on the grain boundaries. Such precipitation can produce zones of reduced corrosion resistance in the immediate vicinity, and it makes the grain boundary zone anodic relative to the remainder of the surface. The attack usually progresses along a narrow path down the grain boundary, and entire grains may be dislodged due to complete deterioration of their boundaries

The morphology of intergranular corrosion is a function of the ratio of the rate of penetration at the grain boundaries to the rate of grain face corrosion. If the ratio is equal to one, there is only general corrosion without any form of intergranular attack. When the ratio is greater than one, there is a high rate of penetration with relatively little attack on the grain face, and the grain faces are resistant to corrosion attack. In this condition, the grains tend to stay in place, and the material specimen appears to be unattacked. However, intergranular attack may have totally penetrated through the entire cross section of the metal specimen and can be readily bent and broken merely by hand. The intergranular corrosion attack proceeds laterally through the sheet in a plane parallel to the surface, and the corrosion products force the metal away from the material and give rise to a leaflike appearance. Such a phenomenon is referred to as exfoliation corrosion.

The major negative effect of intergranular corrosion includes member mechanical properties reduction and impairs the structural member resistance to other forms of corrosion, as demonstrated in the sensitization of stainless steels and weld decay. However, heat treatment and stabilization, barrier coating, and electrochemical protection have proved to be an effective control of intergranular and exfoliation corrosion in engineering facilities.

2.11 Dealloying

Dealloying (selective attack) corrosion is known as preferential removal of one element from an alloy of metals by corrosion processes, leaving the distorted residual structure. This occurs in alloys when one component is more susceptible to attack and corrodes, preferentially leaving a porous material that crumbles. The phenomenon was first reported in 1866 on brass, which consists of copper and zinc alloys. Also, since that time, it has been reported in virtually in all copper-based alloy systems, cast iron, and other alloy systems in the engineering industry.

The dealloying mechanism can be a selective removal of one or more elements that made up of the alloy, leaving a residual substance, or the entire alloy may be even dissolved. A common example is the dezincification of unstabilized brass, whereby a weakened, porous copper structure is produced. Dealloying corrosion can be identified by color changes, such as brasses turning from yellow to red and cast irons changing from silvery gray to dark gray, which may be referred to as graphitic corrosion. During cast-iron graphitization, the porous graphite network is impregnated with insoluble corrosion products, and the cast iron maintains its appearance and shape but is structurally weaker.

The testing and identification of graphitization are accomplished by scratching through the surface with a blade to reveal the crumbling of the iron underneath. The weight loss is not a significant measurement of the impact of dealloying. The depth of attack must be measured by sectioning and microscopic examination. The impact of dealloying on the strength of the material can be known through mechanical testing. The control of dealloying corrosion is generally achieved by prescribing resistant material. Other methods can be effective, such as the use of barrier coatings, reducing the aggressiveness of the environment, cathodic protection, and replacement of the damaged element.

2.12 Stress Corrosion Cracking

Stress corrosion cracking refers to rusting mechanism, which requires the pairing of a material with a corrosive environment and the application of a tensile stress above a critical value. Stress corrosion cracking (SCC) is the cracking induced from the combined influence of tensile stress and a corrosive environment. It is a combination of static tensile stress (external or residual stresses), and corrosion leading to cracks and eventual structure catastrophic failure. In the absence of either the corrosive environment or the tensile stress, cracking may not take place. The diagram showing the stress corrosion cracking controlling factors is presented in figure 2.5.

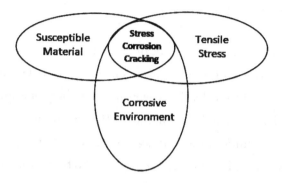

Figure 2.5 Stress corrosion cracking factor

The severity of the cracking increases with the swell of the tensile stress. The increased cracking severity and tensile stress are evident by a greater frequency of failure of a heavily loaded structure in an aggressive environment. The typical morphology of SCC is tight cracks that penetrate the material and are perpendicular to the direction of the maximum tensile stresses. The cracks may be single or multiple cracks, but a group of multiple cracks on metal surfaces is commonly observed. Also, corrosion products may be found in the cracks, and the metal surface may be clean with no evidence of corrosion except for some fine network of cracks. SCC cracks are generally branched and can be propagated along the metal grain boundaries or across the grains.

Stress corrosion cracking fractures have a brittle appearance and are classified as a catastrophic form of corrosion. The detection of such fine cracks can be very difficult, and the damage cannot be easily predicted. However, control of SCC is possible by selecting a material that is not susceptible to the service environment and also by coating to isolate the material from the corrosive environment. Also, controlling of SCC can be done by carefully designing and reducing stress concentrations below the critical value.

2.13 Corrosion Fatigue

Corrosion fatigue occurs as a result of combined action of an alternating tensile stress and a corrosive environment. Corrosion fatigue is simply two different failure mechanisms working together; one mechanism is corrosion and the other is mechanical. This phenomenon has mostly come across on facilities located in aggressive conditions such as offshore environments. The fatigue process is considered to be caused by the rupture of the protective passive film on the component surfaces upon which corrosion is commenced and accelerated. The corrosive environment causes faster crack growth because the resistance of any metal will definitely reduce under corrosive environment. Following Beden et al. (2009), the transition from the corrosion-controlled phase to the fatigue-controlled phase can be characterized by equations 2.1 and 2.2, where ΔK is the fatigue growth and ΔK_{th} represented corrosion growth.

$$\Delta K \geq \Delta K_{th} \qquad (2.1)$$

$$\left(\frac{da}{dt}\right)_{FM} \geq \left(\frac{da}{dt}\right)_{Pit} \qquad (2.2)$$

While the corrosion defect grows, the stress concentration at the tip of the defect increases. The moment the stress intensity ranges ΔK caused by cyclic loading exceeds the threshold ΔK_{th}, fatigue crack growth may start, and final result is failure. The first condition is a precondition for the second condition. The model is simplified, based on the hypothesis that there is no interaction between the chemical (corrosion) and mechanical (stress ranges) deterioration process. The component damaged process is the assumed fatigue controlled, where the fatigue crack growth is considered faster than the corrosion growth.

The fatigue life of a typical structural component is directly linked to the fatigue progression, which can be grouped into three major stages (see fig. 2.6): crack initiation, crack propagation, and fracture failure, as demonstrated by the investigation of Wohler (1893) and Sobczyk and Spencer (1999).

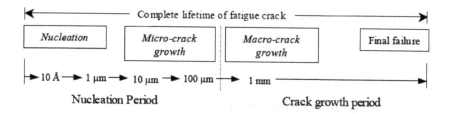

Figure 2.6 Crack nucleation and propagation

Fatigue can be classified as high cycle (low stress) or low cycle (high stress). A fatigue is called low-cycle fatigue if the number of cycles to failure is less than ten thousand. The number of cycles in a high-cycle fatigue is usually several millions. For the marine structures, the latter has been the concern because it is applicable to offshore environment. Identification of a fatigue limit state required knowledge of the mechanical behavior of fatigue process. The fatigue damage progression is an accumulation of material initiated from yielding in the material by the sliding of atomic layers. This is due to repeated load, which may also cause by a combination of displacement and local stress concentrations.

Several microscopic cracks present in the material tend to grow under the influence of recurring loading and later join to each other to form major cracks. The total time of three phases of crack initiation and growth constitute the complete lifetime of fatigue damage accumulation. Based on the material properties and the type of loading, the nucleation phase can be of a different importance in estimating the fatigue life. Collins (1993) concluded in the experimental observations that at high cycle fatigue, a significant proportion of the usable fatigue life may be consumed by the crack initiation period.

The surface quality of material influences the fatigue crack growth, because localized corrosion defects can serve as initial cracks from which crack growth starts. Control of corrosion fatigue can be accomplished by reducing the overall stress level and designing out stress concentrations in the system. Prevention of corrosion fatigue can also be attained by the selection of a suitable material that is not susceptible to the environment. Application of a suitable protective coating is also highly recommended in most circumstances.

2.14 Hydrogen Embrittlement

Hydrogen pressure–induced cracking, also known as stepwise cracking, hydrogen stress cracking, or cathodic stress cracking, results from the action of tensile stress and the ingress of hydrogen into a component. Hydrogen stress cracking is brittle in nature, can be catastrophic, and can occur within hours of exposure to the sour environment.

Cathodic stress cracking is associated with corrosion in corrosion-control processes when atomic hydrogen produced on the metal surface by corrosion-control processes is absorbed by the metal and promotes pressure induced cracking. This phenomenon may seriously reduce the ductility, load-bearing capacity, and catastrophic brittle failures at stresses below the yield stress of susceptible materials.

Hydrogen embrittlement does not affect all metallic materials equally, and the most vulnerable are high-strength steels, titanium alloys, and aluminum alloys. The common causes of hydrogen embrittlement cracking in marine structures are hydrogen produced by rusting and cathodic protection devices.

The factors controlling the risk of hydrogen stress cracking include tensile stress level of applied loads and residual stresses as a result of welding. A component that is subjected to stress relieving heat treatments is less at risk than those in the as-welded condition. No wonder there is a requirement for the thermal stress relief on carbon manganese and low alloy steels intended for sour service if specified hardness limits are not achieved in the as-welded condition. The sulfide stress cracking may occur in hard macrostructures of untampered martensite and other relevant limited hardness material exposed to sour condition. The concentration of hydrogen within the steel components, which is controlled by the pH of the sour environment and concentration of hydrogen sulfide, may also lead to stress corrosion cracking as it is applicable to oilfield environments that contain hydrogen sulfide.

2.15 Microbiologically Influenced Corrosion

Microbiologically influenced corrosion refers to corrosion caused by the activities of microorganisms and their metabolites. Bacteria can change the environment at the steel surface, releasing acid products that cause corrosion. Some bacteria also interact directly with the meal surface and oxidizing the iron. Microbiological corrosion can occur on the metal surface situated in either aerobic or anaerobic environment. That is why bacteria, fungi, and other microorganisms play a major role in the corrosion activities that take place inside the soil, and dramatically rapid corrosion failures have been observed in the structures under the ground due to microbial action.

Microorganisms associated with corrosion damage are classified as anaerobic bacteria, which produce highly corrosive species as part of their metabolism. Also, aerobic bacteria produce corrosive mineral acids and fungi that produce corrosive by-products in their metabolism, such as organic acids. Apart from metals and alloys, which can be degraded by the microorganisms, they may also ruin organic coatings, thereby exposing the metal to corrosion attack.

CHAPTER 3

MARINE CORROSION CHARACTERISTICS

3.0 Introduction

The marine environment is characterized by chlorine particles that get deposited on marine structure surfaces to form part of chemical pollutants, causing corrosion. The marine environment is highly corrosive, and the corrosion rate tends to be significant depending on the water depth, fraction of chlorine in the seawater, and water temperature. Corrosion science is a vital theme in marine oil exploration and engineering. Corrosion is given several challenges to the operators who should guarantee good returns on their investments, as well as to the engineers who must design and protect the structures to withstand the aggressive offshore conditions. Lack of routine maintenance mandates that corrosion control systems should be engineered to a very high level of performance and maximum service life for marine structures.

The major factors that contribute to the hostile environment of marine installations are broadly categorized as follows.

- Continuous exposure to salt spray in atmospheric zones
- Incessant wet/dry conditions from wave action

- Severe exposure to ultraviolet sunlight
- Constant movement and flexing of substrate
- Severe abrasion from drill pipe and casings, boat landings
- Chemical spills and drop objects

The governor elements that control the rate of corrosion in marine environment are the amount of oxygen, water and salt present in the specific zone. The qualitative representation of the above-mentioned elements available in every tidal zone in the offshore environment is presented in figure 3.1. In terms of corrosion locality, the marine environment can be divided into atmospheric zone, splash zone, tidal zone, immersed zone, and seabed sediment zone. The unique features of each marine environmental zone, with radical differences in exposure, cause significant differences in their corrosion progression. Also, the seabed sediment zone is generally considered to be the least corrosiveness region due to the little presence of oxygen.

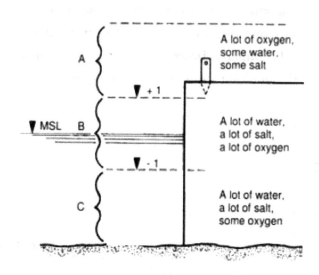

A - Splash Zone, B - Tidal Zone, C - Immersed Zone

Figure 3.1 Marine environment corrosion zones

The maximum severity of corrosion is in the splash zone just above the high tide level and is accounted for by the fact that surfaces in this zone are in continuous contact with highly aerated seawater. The minimum corrosion within the tidal zone is due to the differential aeration cell's protective effect just described. The secondary peak of corrosion just below low tide is also due to the differential aeration cell action in which the surfaces below low tide become anodic to the tidal zone surfaces. This peak is more evident in shallow water because of the smaller area below low tide. Steel structures located in intertidal zone and splash zones are characterized by high corrosion rates. The portions of steel structures that are several meters below the water level on average have lower rates of corrosion because one of the parameters for corrosion process, oxygen, is in short supply underwater.

Jacket member corrosion losses in the splash zone, medium tidal zone, and low tidal zone, based on structural member UT measurement results over the years in the Niger Delta, is revealed in figure 3.2. The graph revealed that the splash zone graph is steeper than other graphs due to accelerated corrosion process in the zone. This phenomenon is a result of high tide and continuous jacket members' contact with aerated warm seawater, which significantly supports corrosion growth.

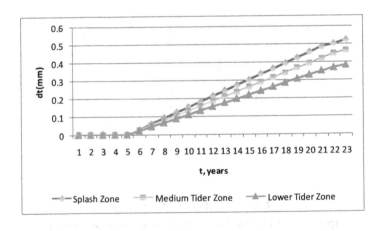

Figure 3.2 Jacket structure corrosion losses versus time

MATTHEW OMOTOSO, PH.D

There are number of factors that influence the rate at which steel structures corrode in seawater. The contributing factor to corrosion process in marine environment is not limited to the factors mentioned previously. Marine steel structure can be corroded due to chemical attack and abrasion. The corrosion process may also increase owing to the presence of dissimilar metals with differing electrical potential. Corrosion process may be accelerated by the action of bacteria even in the absence of oxygen, because sulfate-reducing bacteria produce sulfides that cause rapid attack on steel. Thus, the site of sulfate-reducing bacteria attack is frequently inside steel piles that are depleted in oxygen. The bottom sediment is oxygen deficient, and it is a sulfate-reducing bacterial attack site that is common to the buried steel structure corrosion damaged in sediment. A sulfate-reducing bacteria attack may occur in a polluted harbor with the absence of oxygen. The presence of sulfate-reducing bacteria can often be detected by the odor of rotten eggs produced by the bacteria.

Cathodic protection of marine structures begins from the external protection of the underwater area, including the attachments and opening. For a mooring suction pipe, the part to be considered includes the padeye, mooring chains, vent hatch, and head plates. Passive cathodic protection systems employ sacrificial anodes attached to the protected structures using steel bars or cables. Aluminum anode is popularly used for this purpose due to light weight and low electrical equivalent. Coatings assume the function of passive protection on marine structures and are essential as carriers for antifouling substance. The aim of combining coatings with cathodic protection is to reduce the protection current requirement and increase the protection rage by raising the polarization parameters. Nowadays, marine structures are providing additional thickness of steel as corrosion allowance to compensate for corroded metal thickness that may take place during the facility design life. Identified corrosion allowance is usually of the order of three to twelve millimeters depending on the material, design specifications, and applicable codes.

3.1 Seawater Resistivity

The extent of material interaction in the corrosion process is partially dependent on the resistivity of surrounding medium. A medium with low resistivity, such as seawater, provides a greater degree of electrical connectivity than medium or higher resistivity. The resistivity of seawater varies with the amount of salt content and temperature. Salt content may also vary for many reasons, including an influx of fresh water or evaporation. The following are the major seawater parameters that dictate the degree and extent of corrosion damages of marine installations: seawater salinity, oxygen content, seawater current, water temperature, and marine growth. The previously mentioned parameters are therefore considered based on the environmental data in the course of corrosion control systems design as it is applicable to cathodic protection systems.

3.2 Seawater Salinity

The salt content of seawater categorizes the waters in which marine structures are built and the waters through which ships travel. The approximately typical seawater salinity is between 3 percent and 4 percent, and they are mostly sodium chloride. The salt content determines the specific electrical conductivity of the water, and it is particularly important because they have profound influence on the action of corrosion cell. In strong sunlight, water can evaporate a defective coating and surface films on marine structures, leading to concentration and crystallization of salts and giving rise to local anodes that may damage the surface films.

3.3 Dissolved Oxygen Content

Oxygen role in seawater corrosion is a complex process. With some materials such as steel, it accelerates corrosion by serving as a cathodic depolarizer. However, with stainless steel, oxygen retards corrosion by the development and repair of oxide films that are responsible for the passivity of the metal. Oxygen is an essential element in the cathodic partial reactions, and variations in oxygen concentration on a metal surface generate corrosion currents.

The surfaces with the smallest oxygen concentration suffer accelerated corrosion as anodes in a differential aeration or oxygen concentration cell. Such a cell can be developed at the mudline with the higher oxygen content in the water and the lower content below the mudline. Consequently, the cell operates to remove metal below the mudline. This is also applicable to local corrosion observed on the boat landing of an offshore jacket platform due to differential aeration leading to shallow pitting corrosion in some cases.

3.4 Sea Current

Seawater flow rate raises the rate of oxygen transportation and also adversely affects surface films formation on steel members. The velocity of seawater flowing past marine structures is known to increase the rate of marine immersion corrosion. The effect is most significant in the early corrosion period before corrosion products formation and marine growth, because this elements offer protection from the effect of sea current. Offshore structures splash zone is the most critical area for corrosion due to waves leaving an oxygen-saturated film on the metal surface in conjunction with mechanical action disrupting the layers of semiprotective corrosion products.

3.5 Temperature

In principle, differences in water temperature may lead to the formation of differential temperature cells. However, this phenomenon may have little effect below the waterline. Oxygen solubility decreases almost linearly with increasing temperature, but the diffusion rate increases exponentially.

Therefore, an increase in water temperature results in a slight increase in corrosion rate. As a result, greater corrosion rate is considered in tropical regions like Nigeria waters compared with the North Sea.

3.6 Marine Growth

Marine life such as algae, barnacles, and mussels has been the scourge of marine installation for centuries. Marine growth accommodates bacteria, which produce highly corrosive minerals that corrode metal and degrade coatings. That makes corrosion attacks on marine structures become an emergency issue that requires urgent attention. Marine growth is an issue for offshore operators and engineers because it is vitally important to prevent biofouling on offshore installation. Excessive thickness of marine growth on offshore installations creates added area for wave loading and drastically reduces the fatigue life of offshore installations. Excess marine growth becomes an issue because weight increases the strength requirements of a structure.

3.7 Soil Corrosion Factor

Soils are an admixture of rock disintegrated by physical action and modified by weathering, materials precipitated chemically from aqueous solution, sediments, and organic matter. There are many different forms of soils, and each is recognized and categorized according to its texture and color. Coarse and loose soils allow free circulation of air, and therefore the corrosion process in this soil may be similar to atmospheric corrosion. Soil with fine texture retains more water, leading to poor aeration and drainage, which supports an accelerated corrosion process relative to loose soil.

The common method of choice for transportation of water, natural gas, oil, and refined hydrocarbons remains that of buried pipelines. All over the world, there are thousands of kilometers of buried petroleum products pipelines, which contribute billions of dollars to the global economy.

The replacement value for such facilities represents billions of dollars. This mode of transportation remains the safest way to ship oil and gas, but deterioration of line pipe steel due to external corrosion can lead to stunning failures. The major soil characteristics that dictate the extent of corrosion damages of buried steel installations are soil category, moisture content, water table, soil resistivity, and soil pH.

3.8 Soil Category

Soil structures and particle size arrangement determine the physical properties of soil matrix and permeability. Soil permeability in turn controls the rate of movement of fluids and gases through the matrix. Thus, soils made up of a broad distribution of small particles like clay are preventive, whereas well-sorted coarse sands allow greater oxygen flow. When the oxygen content of a soil is high, the oxidation-reduction potential will be more positive than that soil with less oxygen. The oxidation-reduction potential is somewhat of an abstract value, but field evidence indicates that this parameter may be used for the prediction of soil corrosivity.

Capillary force in low permeability soils also draws groundwater to form a water-saturated zone and raise the local water table. Fine aggregate soil like clay may exert physical force on underground structures by sticking to coated steel surfaces as a result of the soil swelling and shrinking, which is caused by hydration effects, damages the coating, and provides opportunity for corrosion attacks.

3.9 Soil Moisture Content

The soil moisture content is determined by the amount of rainfall and the water table. The water table is defined as the top of a water-saturated zone where water forms the continuous phase in the soil matrix. The water level rises and falls seasonally with the rate of precipitation and capillary effects in the soil matrix. The existence of moisture in the soil is a key requirement for the corrosion of buried structures. Dry soils are not of concern with respect to corrosion. The position of the water table likewise influences the nature of the corrosion process because it determines rates of oxygen transport in the soil.

3.10 Soil Resistivity

Soil resistivity is a measure of how much the soil resists the flow of electricity. The resulting soil resistivity is expressed in ohm-meter or ohm-centimeter. Soil resistivity testing is the single most critical factor in electrical grounding design. An understanding of the soil resistivity and how it varies with depth in the soil is required to design impressed cathodic protection systems. The soil resistivity decreases as soil moisture content increases. The soil resistivity may vary widely with depth below the ground surface. Generally, there is a belief that when soil resistivity becomes lower, corrosion of unprotected buried metal becomes faster. Also, cathodic protection system becomes more effective as resistivity of soil becomes lower. The soluble ion content of the soil determines the resistivity of the soil. The increased soluble ion content decreases soil resistivity, and unprotected metals' corrosion rate increases. Certain ions like calcium and magnesium ions will behave differently because they may form insoluble carbonate deposits at the metal surface, therefore lowering the metal corrosion rate.

The most common indirect technique of determining the soil resistance is the Wenner method, or four-pin methods, using commercial ohmmeters with four-connection clamps, as displayed in figure 3.3. When using the four-pin method, the four pins should drive into the ground in a straight line at the preferred spacing. A good contact to the soil is required for the accurate measurement. The two C binding posts are connected to the endpins, and the two P binding posts are connected to the adjacent center pins.

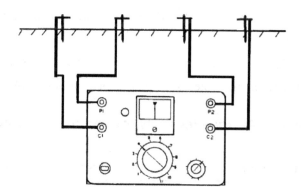

Figure 3.3 Four-pin methods

The resistivity measurements provide a part of data needed by corrosion engineers to design and operate a cathodic protection system. The specific soil resistivity is determined by using equation 3.1, where I denote the injected current between electrodes A and B, U is voltage between the electrodes C and D. The distance between the pins is denoted with "a"

$$\rho = 2\pi a \frac{U}{I} = 2\pi a R \qquad (3.1)$$

The direct method of measuring specific soil resistivity is carried out in the laboratory on soil sample using a soil box. The specific soil resistivity's value is determined by using equation 3.2. In both cases, the measurement may be made with AC to suppress polarization effects. While making measurements in a box, it should be noted that soil samples could transform from original condition and make the resistivity value obtained inaccurate. Therefore, soil resistivity measurements in the soil box give only accurate results with cohesive soil using equation 3.2.

$$\rho = R\frac{S}{L} = \frac{U \times S}{I \times L} \qquad (3.2)$$

L is soil specimen length, S is the soil specimen cross section, and U in the applied voltage.

Investigation of resistivity for different types of soil and seawater in the Niger Delta area for the project's purpose is presented in table 3.1

Table 3.1 Average resistivity of some soil and water

Environment	Resistivity Range (ohm/cm)
Seawater	25.5
Alluvial Soil	1,000–2,000
Clays	1,000–5,000
Gravel	10,000–25,000
Sand	25,000–50,000

3.11 Soil pH

The pH of soil is usually within the range of 3.5–10.0. Soils contain organic matters that are acidic due to the leaching of different type of cations by rainwater, in addition to dissolving of carbon dioxide into the groundwater. The corrosion of iron as a function of pH increases considerably at pH values less than 4 because the metal passivation occurs at high pH values. However, in contrast to iron metals, aluminum may be rapidly corroded in alkaline soils with less pH values as well as in acidic environments.

3.12 Soil Microbes

Microbial-influenced corrosion can be defined as the deterioration of metals by natural processes directly or indirectly related to the activity of microorganisms. There are many types of microorganisms involved in the microbial corrosion that are responsible for the corrosion failure of mild steel, stainless steel, and copper and its alloys. Sulfate-reducing bacteria is considered one of the well-known anaerobic bacteria responsible for severe corrosion damage when observed under anaerobic conditions in wet and black soil. The bacteria organisms reduce the sulfate to sulfide. The attack of sulfate-reducing bacteria is easily detectable by forming iron sulfide (FeS) with black film. The microorganisms associated with corrosion damage are classified as aerobic bacteria that produce highly corrosive species and corrosive mineral acid.

The most damaging corrosion takes place in the presence of a multispecies biofilm. In such biofilms, the interactions between different species may induce a cascade of biochemical reactions in the oxic and anoxic parts of the biofilm and exacerbate corrosion (Videla and Herrera 2009). Multiple corrosion-causing organisms in a biofilm can act synergistic and contribute to more severe corrosion than when only one single species is present.

CHAPTER 4

CORROSION MODEL

4.0 Introduction

A complete corrosion model should consist of modeling both temporal and spatial variability. A proper corrosion model should be able to predict the rusting spatial distribution and depth at a specific point in time. However, most corrosion models that are available for use are limited to prediction of rusting temporal characteristic processes. The past studies simply consist of data collection without detailed analysis, which results in statistical models leading to a degree of uncertainty, low predictive capability, and little practical use for decision-making processes. Therefore, to understand the influences of various factors, it is necessary to move away from observations but to study corrosion behavior.

Deterioration of engineering facility components is often due to aging effects, corrosion, and time-dependent failure mechanisms, which can manifest itself on a continuous basis and seldom superimpose on each other. Marine corrosion of steel is influenced by a variety of factors. The factor includes contact with seawater, temperature, variable wetting and drying cycles, access to oxygen, marine organism fouling, bacterial action, wave impact, currents, and geographic location. Most of these factors are time dependent, and their consequences on the marine corrosion have different rate and intensity.

4.1 Corrosion Growth Model

Despite the widely known corrosion damage of carbon steel in the petroleum industry, mild steel is still used in large tonnages in construction applications worldwide. Hence, for time-dependent reliability assessment of structures, the understanding of the factors affecting corrosion process and the development of corrosion models, which may efficiently forecast long-term corrosion wastage, is of major importance. The geometrical characteristics of corrosion defects are the first approach to the complex corrosion problem, which facilitates a stochastic description of the defects. Corrosion phenomena can be distinguished by their geometrical characteristics without considering their driving mechanisms. The total corrosion depth at any location *(x)* and time *(t)* can be described by the sum of the two types of defects, as presented in equation 4.1.

$$d_C\left(x,t\right) = d_{UC}\left(t\right) + d_{LC}\left(x,t\right)$$

(4.1)

In this equation, $d_C(x,t)$ is equal to the depth of corrosion at location x at the time t, $d_{UC}(t)$ is the depth of the uniform corrosion, and $d_{LC}(x,t)$ is the depth of the localized or pitting corrosion defect. Steel member corrosion losses may lead to structure failure due to structural resistance reduction. The reliability calculations based on limit state have been presented for ship structures with generic form of limit state function for structural member corrosion loss is represented in Equation 4.2 as proposed by Yong Bai (2003).

$$g_C = d - d_{crit} - d_{UC}$$

(4.2)

The depth d_{UC} of the uniform corrosion which is applied for the reason that uniform corrosion normally has less influence on the structural resistance, d_{crit} is the critical member thickness loss at which failure occurs.

From a mechanism point of view, seawater corrosion on steel structural elements may be generally associated with widespread decrease of thickness or may be localized, resulting in deep holes. It is generally known that the corrosion evolution process in the marine environment consists of different corrosion phases. The first of these dominates at the beginning and is characterized by continuous coating breakdown and a decrease of thickness of all exposed steel surfaces, which may be referred to as uniform corrosion. As time continues, some parts of the surfaces still experience continuous loss of thickness, which may be called localized corrosion. The initiation of pitting corrosion may arise, and this phenomenon becomes more deepened with time. The final phase is severe material deterioration due to pitting holes growth that can lead to structural failure. A schematic representation of these phases is presented in figure 4.1.

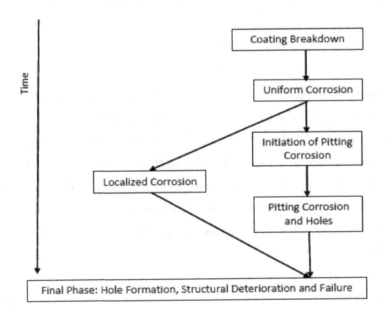

Figure 4.1 Schematic representation of pitting corrosion model

4.2 Time to Corrosion Damage

The period required for steel structures in marine environments to experience corrosion damage is equal to the time needed for the chlorine ion to diffuse down to the steel component surface and accumulate in concentration in excess of the corrosion threshold, as well as the time for corrosion to occur and generate enough rust material to fall off from the steel component surfaces. The rate of marine steel component corrosion losses is directly proportional to the amount of rust generated and the later falling off from the member surfaces. The rate of production of rust is determined using an equation from Youping Liu (1998).

$$\Delta S^2_{crit} = 2 \int_0^t 2.59 * 10^{-6} \left(\frac{1}{\alpha}\right) \pi D i_{corr}(t) dt \qquad (4.3)$$

Here, α is the molecular weight of steel or corrosion products (mg), $i_{corr}(t)$ is the rate of corrosion as a function of time (mm/yr), t is time in years, and the and ΔS_{crit} is the critical volume of corrosion product required to fall off (mm³).

The time required to generate the volume of rust is obtained by applying equation 4.2, but this model was validated in the laboratory and therefore is inappropriate to be used for field facilities because the process rate varies with time and other factors.

Engineering practice, supported by field experience, suggests that measuring the corrosion rate several times and adopting the average corrosion rates measured is the actual corrosion rate. This method is frequently applied in the petroleum industry by using NDT equipment to measure marine steel structure thickness during facility inspection.

The period required for steel components to experience corrosion losses can be represented mathematically, as described in equations 4.4 and 4.5 for uncoated and coated marine steel structures, respectively.

$$T_d = T_i + t_p \tag{4.4}$$

$$T_{d1} = \tau + T_i + t_p \tag{4.5}$$

where T_d is time for corrosion losses, T_i is time for chlorine ion accumulation in excess of corrosion threshold, t_p is time for corrosion occurrence and rust material falling off, τ is coating life span, and T_{d1} is time for corrosion losses for coating component.

4.3 Salinity and Marine Corrosion

Modeling durability of marine structures due to corrosion damage requires quantitative understanding of the structure environment and structural component physical deterioration processes. Equations for each part of these processes are available. However, several models that are available in the literature have been developed in a particular environment that was not suitable for an offshore oil production environment associated with accidental ion discharge, ocean waves, and current action.

The chlorine ion is one of the agents responsible for corrosion in a marine environment. Coated steel develops a passive oxide layer that is highly protective, and corrosion grows at a very slow rate. Seawater typically contains high percentages of sodium chloride, although the salinity may be higher in some areas with accidental discharge of chlorine substance. The rate of corrosion is controlled by the chloride content and availability of oxygen.

The process of chloride-induced corrosion for marine steel structure is by diffusion of chloride through the damaged coating. The chloride builds up over time on the component surfaces until attaining a critical threshold; then the passive oxide layer on the steel breaks down, and corrosion begins. The replacement of corroded component may be made. However, the cycle continues on the new component as well. Most models assume that the dominant process is diffused for a practically well-constructed structure with good coating quality. Diffusion calculation is a reasonable approximation of the overall real process for chlorine ion transportation. The diffusion process is modeled by solving one dimensional equation for Fick's second law of diffusion.

$$\frac{\partial C}{\partial t} = D\left(\frac{\partial^2 C}{\partial x^2}\right)$$

(4.6)

where C is salt ion concentration (mol/m³), t is time (seconds), D is diffusion coefficient (m²/s), and X is the position (length, meters).

Equation 4.7 can be derived from Fick's first law and the mass balance.

$$\frac{\partial C}{\partial t} = D\frac{\partial}{\partial x}J = \frac{\partial}{\partial x}\left(D\frac{\partial}{\partial x}C\right)$$

(4.7)

If the diffusion coefficient D is constant, we can exchange the orders of the differentiating and multiplying by the constant.

$$\frac{\partial}{\partial x}\left(D\frac{\partial}{\partial x}C\right) = D\frac{\partial}{\partial x}\frac{\partial}{\partial x}C = D\frac{\partial^2 C}{\partial x^2}$$

(4.8)

In the case of diffusion in two or more dimensions, Fick's second law becomes

$$\frac{\partial C}{\partial t} = D\nabla^2 C$$

(4.9)

In a condition in which concentration does not change from time, the above equation becomes zero, which is Laplace's equation. This equation is usually solved using the error function solution.

$$C_{(x\,t)} = C\left[-erf\left(\frac{}{\sqrt{Dt}} \right) \right] \qquad\qquad (4.10)$$

where $C_{(x,\,t)}$ is salt concentration at depth x and time t, C_0 is surface concentration (%), *erf* is the error function, and ∇ is used for two or more dimensions.

4.4 Chloride Ions Concentration Growth

Fick's second law of diffusion, discussed in the previous sections, specified that chloride ion concentration (C_0) has to remain constant during the course of the diffusion process. However, the surroundings of offshore production facilities are characterized by accidental discharge of ions that make chlorine ion increase with time. The chlorine ion concentration of the seawater around the facility may also be known via chemical analysis in the laboratory.

Fick's second law predicts how diffusion causes the concentration to change with time. To account for the increasing C_0, it was assumed that C_0 increases linearly, satisfying equation 4.11.

$$C_O = mt \qquad\qquad (4.11)$$

Here, m is the rate of chloride accumulation (mol/m^3/yr), and t represents the age of the facility (in years). The solution to Fick's second law in equation 4.10 is developed further in equation 4.12 to account for the increase in chlorine ion concentration around the platform.

$$C_{(x,t)} = \int_0^t m\left[1 - erf\left(\frac{x}{2\sqrt{Dt}} \right)\right] dt \qquad\qquad (4.12)$$

4.5 Advection Term and Diffusion Equation

Advection is the component of solute movement attributed to transport by the flowing water. The rate of transport is equal to the average linear water velocity. Equation 4.11 is further developed to account for the ocean waves and current that are constantly instigating water movement in the surrounding of the platform. As seawater moves, the concentration of the chlorine ion in the seawater is affected by both physical processes and diffusion processes that make it necessary to introduce the advection term to account, as expressed in equation 4.13.

$$\frac{\partial C}{\partial t} = D\left(\frac{\partial^2 C}{\partial x^2}\right) - v\left(\frac{\partial C}{\partial x}\right) \tag{4.13}$$

The boundary conditions represented by the step-function input are described mathematically as follows.

$$C(X,0) = 0 \qquad x \geq 0$$
$$C(0,t) = 0 \qquad t \geq 0$$
$$C(\infty,t) = 0 \qquad t \geq 0$$

For the above boundary conditions, the solution to equation 4.13 for a saturated homogeneous porous medium is found in Ogata (1970).

$$\frac{C_O}{C_1} = \frac{1}{2}\left[\text{erfc}\left(\frac{x - vt}{2\sqrt{Dt}}\right) + \exp\left(\frac{vx}{D}\right)\text{erfc}\left(\frac{x + vt}{2\sqrt{Dt}}\right)\right] \tag{4.14}$$

Here, *erfc* represents the complementary error function, x is the distance along the flow path, and v is the average linear water velocity. For the conditions in which the dispersity of porous medium is large, the second term on the right-hand side of equation 4.14 is negligible. The effect of longitudinal dispersion is illustrated by a simple column experiment for a step-function input, and a concentration profile described by Freeze and Cherry (1979) is adopted in equation 4.15.

$$\frac{C_O}{C_1} = \frac{1}{2} \text{erfc}\left(\frac{x-vt}{2\sqrt{Dt}}\right)$$

(4.15)

This initial problem in equation 4.12 may be expressed in simplified form in equation 4.16.

$$C_{(x,t)} = \int_0^t m\left[1 - erf\left(\frac{x}{2\sqrt{Dt}}\right)\right] dt - \frac{1}{2} erf\left(\frac{x-vt}{2\sqrt{Dt}}\right) dt$$

(4.16)

4.6 Chlorine Ion Concentration and Accumulation

The Metocean data in Nigeria Niger Delta and laboratory analytical of chlorine-ion concentration in the vicinity of some offshore platforms were presented in tables 4.1 and 4.2, respectively. Table 4.2 revealed that chlorine ion concentration in the immediate surroundings of crude oil offshore production facilities, which is 3.7 mole/m^3 was higher than the open seawater, which is normally be 3.5 mole/m^3. This occurrence was due to the release of chlorine ion to offshore platform surrounding during crude oil production and offloading activities.

Table 4.1 Niger Delta Metocean data (Santala 2002)

Metocean Directions (0o)	Waves Period (s)	Wave Velocity (m/s)	Wave Length (m)
20	14.0	0.41	4.34
65	13.0	0.57	7.41
110	5.0	0.57	2.85
155	5.0	0.31	1.55
200	5.0	0.41	2.05
245	5.0	0.57	2.85
290	5.0	0.57	2.85
335	13.5	0.36	4.86

The production platform in table 4.2 was used as a case study with a chlorine ion concentration of 3.7 mole/m^3 at 20° Metocean direction, and the platform was installed twenty-one years ago. Based on Metocean data in table 4.1 and the application of equations 4.10 and 4.15, the chlorine accumulation rate and chlorine ion concentration based on Metocean direction were determined and presented in table 4.2. Table 4.3 revealed that the rate of chlorine-ion concentration varies according to Metocean directions due to differences in wave velocity despite the fact that the rate of chloride accumulation of 1.8 mol/m^3/yr were the same in the surrounding of production platform.

Table 4.2 Facility age and chlorine ion concentration in Niger Delta

Facilities Description	Facility Age (yr)	Metocean Directions (0o)	Wave Velocity (m/s)	Chlorine Ion Concentration (mol/m3)
Production Platform	21	20	0.41	3.70
Offloading Buoy	20	20	0.41	3.59
Wellhead Platform	20	20	0.41	3.58
Processing Platform	20	20	0.41	3.51
Living Quarters Platform	20	20	0.41	3.62

Table 4.3 Chlorine-ion concentration around production platform

Metocean Directions (0o)	Platform Age (yr)	Average Wave Velocity (m)	Chlorine-Ion Concentration (mol/m3)	Average Chloride Accumulation (mol/m3/yr)
20	21	0.41	3.7	1.8
65	21	0.57	3.7	1.8
110	21	0.57	3.7	1.8
155	21	0.31	3.8	1.8
200	21	0.41	3.7	1.8
245	21	0.57	3.7	1.8
290	21	0.57	3.7	1.8
335	21	0.36	3.7	1.8

The graph in figure 4.2 shows the relationship of the chlorine ion versus ocean wave velocity along Metocean directions. Also, figure 4.3 reveals the trend of chloride accumulation versus chlorine ion concentration. The sketch shows that the rate of chlorine ion concentration is directly proportional to the rate of chloride accumulation. Likewise, figure 4.4 shows that chlorine ion concentration is inversely proportional to ocean wave velocity. As the wave increases, the rate of chloride accumulation decreases, and vice visa. However, the value of chlorine ion concentration in the vicinity of production platform shall not be less than 3.5 percent, which is the chlorine ion concentration value in the open seawater whatsoever the magnitude of the ocean wave velocity.

The sketch in figures 4.3 and 4.4 was combined to produce the sketch in figure 4.5. The drawing shows an open seawater chlorine ion concentration of 3.5 percent. But as chlorine accumulation increase s, the value of chlorine ion concentration increases. Similarly, as ocean wave velocity increases, the chlorine ion concentration decreases. The intersection of the two slope graphs with chlorine ion concentration of 3.8 percent may be defined as chlorine ion concentration equilibrium in the vicinity of the case study production platform.

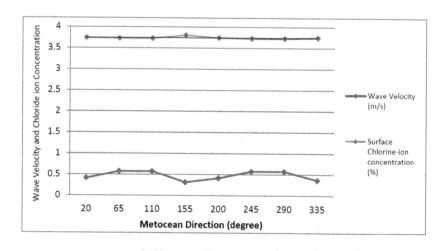

Figure 4.2 Graph showing the effect of chlorine ion versus ocean wave velocity

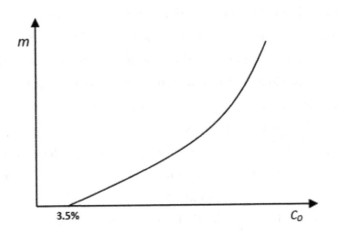

Figure 4.3 Indicate Chloride Accumulations (m) Versus Chlorine-ion Concentration (Co)

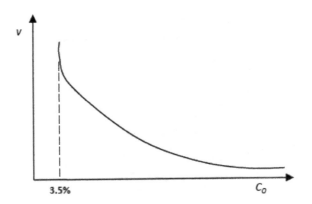

Figure 4.4 Indicate Seawater Velocity (v) Vs Seawater Chlorine-ion Concentration (C_o)

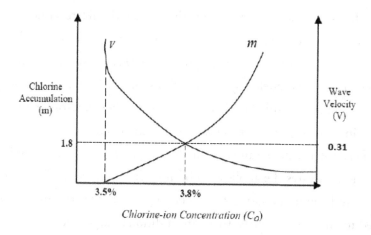

Figure 4.5 Indicate ocean wave velocity, chlorine accumulation, and chlorine ion concentration

4.7 Uncertainty in Corrosion Model

The quantitative corrosion models being applied by engineers are, by and large, developed for design purposes. These represent a sort of worst-case model without clear definition of underlying uncertainties. The absence of uncertainties in any existing model constitutes a significant drawback in the completeness of many corrosion models. Most of the models suggested so far take basis on the assumption that the rate of corrosion is constant. This may be evaluated by a time average of degradation measurements over a sufficiently long interval. Furthermore, most published works assume that the condition assessment of facilities can be based on models that represent the conditions at only one given location. It is obvious that these approaches disregard the aspects of both temporal and spatial variations of the corrosion processes.

The quantitative corrosion models that are available in some of the literatures predict the extent of corrosion loss at a specific location and time as a function of the environment and the material. But the models are highly simple and do not explicitly address the related uncertainties due to limited understanding of the underlying processes and variability in the behavior of corrosion. Also, marine-immersed steel specimen corrosion behavior has been modeled by many authors, but such a phenomenological modeling may be difficult to achieve in the other environment due to several influencing parameters that are not homogenous.

Therefore, the uncertainty in quantitative corrosion predictions is very large and motivates the application of in-service inspections for corrosion control and model updating by the engineers. Besides the uncertainties related to corrosion models, there are additional large uncertainties with regard to environmental and operational conditions that determine the input parameters to corrosion models. The statistical models based on data collection exercises sometimes yield poor quality information and high statistical uncertainties. Therefore, a functional corrosion model cannot be developed without the aid of a good understanding of the corrosion processes in particular environment and conditions.

CHAPTER 5

CORROSION MONITORING TECHNIQUES

5.0 Introduction

Corrosion monitoring is a method that appraises and monitors equipment components, structures, process units, and facilities for any signs of corrosion. Monitoring programs aim to detect asset threaten conditions that can lead to poor performance, safety issues, and unexpected shutdown. The key advantage for implementing corrosion monitoring is to detect early warning signs, corrosion trends, and processing parameters that induce corrosive environment. Similarly, corrosion monitoring serves to measure the effectiveness of corrosion prevention methods and whether different inspection monitoring techniques should be employed.

Corrosion has been established as the predominant factor causing widespread, premature deterioration of construction material worldwide, particularly of the installations located in marine environments. Corrosion monitoring plays a vital role throughout the equipment lifecycle. Inspection and monitoring strategies and techniques may change according to the age and condition of equipment. Therefore, inspection and monitoring strategies should be reviewed at specific intervals at the owner-operator's discretion. In the current business environment, successful enterprises cannot tolerate major corrosion failures, particularly the one that involves unscheduled shutdowns, environmental contamination, personnel injuries, and fatalities. However, a considerable catalyst for the advances in corrosion inspection and monitoring technology has been the exploitation of operator resources in extreme conditions, such as a marine environment with several corrosion agents. Therefore, considerable effort has been expended on corrosion control both at the design and operational stages. Corrosion monitoring and control at the operation phase is very significant, particularly for aging structures, many of which may operate beyond the design life. The corrosion inspection and monitoring flowchart in figure 5.1 is highly recommended for facilities in offshore environment.

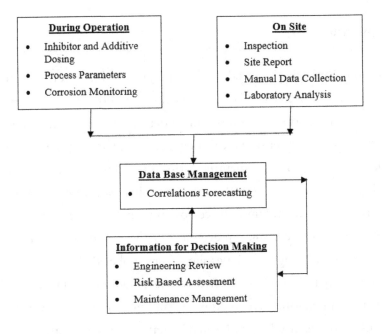

Figure 5.1 Corrosion inspections and monitoring flowchart

The maintenance culture of engineering facilities depends on the severity of the operating environment and the criticality of the engineering system. Some facilities may require only regular cleaning and repainting, but chemical-processing plants and marine structures require extensive inspection and maintenance schedules. Therefore, corrosion inspection and monitoring are used to determine the condition of a system, particularly to ascertain how well corrosion control and maintenance programs are performing. The corrosion-monitoring techniques include simple exposure of coupons, UT test, and smart structure computerized sensing systems. Corrosion inspection and monitoring are beneficial and cost-effective when they are utilized in an integrated manner, and they are complementary and should not be viewed as substitutes for each other (Roberge 1999). How data from various sources should be combined to ultimately produce management information for decision making is illustrated in figure 5.1. These are in three stages as follows: operation data, database management system, and information for decision making.

The significance of corrosion monitoring in engineering system is enormous. Experience shows it may be difficult for a corrosion engineer to get management commitment to provide funds for such programs as due. However, the benefits that can be obtained from investing on timely corrosion monitoring and inspection can hardly be overemphasized, and they include the following.

- Improved personnel and engineering systems safety
- Reduction in operation downtime
- Early warning prior to the occurrence of serious damage
- Reduction in maintenance cost
- Lessening pollution and contamination hazard
- Decrease in operating cost
- Increased life span of engineering installation

5.1 Development of Corrosion Monitoring Program

Corrosion engineers and inspectors should be well aware of how initial considerations of corrosion concerns could be extended to provide a basic corrosion inspection program. Inspection and monitoring techniques and their limitations dictate the best methods to be employed for a specific plant. A number of factors have been identified for essential corrosion monitoring programs in order to make useful contributions toward corrosion control and management. There are several technical requirements identified for the successful corrosion monitoring. Some of these factors include selection of proper corrosion techniques, identification of appropriate location for corrosion monitoring points, data analysis, interpretation, and presentation of the result in a clear and unambiguous format.

The establishment of a confidence factor through experience and correlation with other data, such as failure analysis and plant operational considerations, is very important in corrosion management. The availability of specialist corrosion personnel and infrastructure with adequate budgets to support monitoring programs are is essential. Corrosion control commences at the design stage, and it is essential that consideration is given to feedback from existing corrosion monitoring programs. The data from previous monitoring operations to designers is apparently vital for design improvements. In practice, corrosion monitoring is generally considered to be an enhancement to the usual inspection techniques, but not a replacement.

As soon as severe corrosion crisis is identified through inspection, a corrosion-monitoring program should be launched to investigate the problem in greater depth. A leader of a corrosion-monitoring program is expected to function effectively and efficiently in a multidisciplinary environment and corrosion-monitoring systems. He should be able to engage in a wide range of functions in an organization, such as design, operations, inspection, and maintenance team. To facilitate effective communication and involvement of management in corrosion issues, corrosion-monitoring

MATTHEW OMOTOSO, PH.D

data have to be processed into information suitable for management and nonspecialist understanding. However, modern computing technology can be used to achieve the previously mentioned targets. The summary of commonly used techniques for corrosion monitoring and their applicability are outlined in tables 5.1 and 5.2.

Table 5.1 Corrosion monitoring and inspection methods

S/N	Techniques	Applications
1	**Acoustic Emission**	Measures the location, initiation, and propagation of cracks and defects under stress in metals
2	**Dye Penetrants**	Procedure used for locating surface cracks
3	**Magnetic Particle**	Surface and subsurface crack and defect location, seams, and inclusions
4	**Radiography**	X-ray and gamma-ray measurement of wall thickness, crack, and location of defect
5	**Thermography**	Identifies the lack of bond, hot spots, local thinning, and temperature changes due to poor lagging
6	**Ultrasonic**	Specify internal defects, porosity, lack of fusion, crack location, and wall thickness
7	**Visual Examination**	Localized corrosion identification, erosion, pitting, deposit scaling and fouling problems, staining, and corrosion leakage due to cracking.

Table 5.2 In-plant corrosion monitoring and inspection methods

S/N	Techniques	Applications
1	**Weight Loss Coupons**	Traditional method of limited sensitivity but used in all environments. Type of corrosion observed is an important indicator.
2	**Electrical Resistance (ER)**	Measures change in probe resistance. Widely used for carbon steel fabrications in gas and liquid phases.
3	**Linear Polarization Resistance**	An electrochemical DC method used for measurement of uniform corrosion. Standard electrochemical technique requires a conductive electrolyte.
4	**Zero Resistance Ammetry**	Established method for assessing galvanic corrosion between dissimilar metals.
5	**Hydrogen Probes**	Measure rate of diffusion of hydrogen through steels, either by means of a pressure gauge or by electrochemical techniques.
6	**Thin Layer Activation**	Measures change of radioactivity as a local irradiated area corrodes.
7	**Electrochemical Impedance**	An AC method used for general corrosion measurements; similar to LPRM. More versatile and accurate than DC measurements.
8	**Electrochemical Noise**	A more recent technique used for assessing general corrosion and current and potential fluctuations associated with localized corrosion.

5.2 Corrosion-Monitoring Methods

The corrosion-monitoring and inspection techniques range from visual examination, simple coupon exposures, and nondestructive evaluation to complete engineering facility system surveillance units with remote data access. The conditional diagnosis of engineering installation makes it possible to characterize corrosion origin and the extent of possible damage. In practice, corrosion diagnosis is carried out after some damages have appeared. In general, corrosion diagnosis has several objectives, which include the following.

- Identify corrosion origin and forms
- Evaluate the corrosion damage extent
- Evaluate different materials and their metallurgy
- Appraise different corrosion protection and device performance
- Provide management tools for asset integrity management
- Estimate corrosion damage consequences with regard to safety
- Define further corrosion preventive actions

As soon as corrosion activities are detected, the rate and origin should be determined to sort out which control measure to be applied, or alternatively whether the corrosion damage parts should be totally replaced. However, in choosing any corrosion monitoring techniques, due consideration of the following important factors should be reviewed with financial department and facilities operator.

- Cost
- Safety
- Production
- Ecological concerns
- Public relations

Corrosion-monitoring techniques can be basically divided into two categories: direct and indirect methods. The direct techniques measure parameters, which are directly associated with the corrosion process, such as the measurement of potentials and current flow associated with corrosion reactions. Indirect techniques measure the parameters that are indirectly related to corrosion damage. An example of this method is the measurement of corrosion potential, which is indirectly related to the severity of corrosion damage.

Another classification of corrosion monitoring method is either an intrusive or nonintrusive technique. Intrusive technique is a direct access to the corrosion environment, such as pipeline damage and similar structures by applying sensors and test specimens. Nonintrusive is the opposite of the intrusive method; there is no application of hardware to perform corrosion measurement. Additional corrosion measurement procedures are online and offline corrosion measurement techniques. The online techniques are those methods that require constant monitoring during operation. Off-line methods necessitate only occasional sampling and analysis. Several sections below shall briefly describe some corrosion monitoring and inspection techniques with basic selection principles for the reader understanding.

5.3 Metal Coupons Testing

Corrosion coupons are preweighed and measured metal strips that are mounted in a special pipe system called a coupon rack. They are used to estimate the rate of metal corrosion by comparing the initial weight with the weight of exposure to the fluid in the system. This is the simplest and oldest-established method of estimating corrosion losses in plant and equipment based on weight loss examination. A weighed sample (coupon) of metal or alloy under consideration is introduced into the process, and later it is removed after a reasonable time interval. The coupon is then cleaned of all corrosion products and is reweighed. The weight loss is converted to a total thickness loss, or average corrosion rate, using appropriate mathematical conversion equations.

The metal coupon is simple and typically of low cost. Many forms of corrosion can be monitored if detailed analysis is performed after the outcome of the exposure. However, long exposure periods may be required to obtain significant and measurable weight-loss data. The cleaning, weighing, and microscopic examination of coupons is usually labor-intensive. Corrosion coupon analysis involves a number of variables that may significantly affect the results of the analysis. Make sure the design of the system and the installation of the coupon rack will produce accurate test results.

5.4 Electrical Resistance Measurement

The electrical resistance technique is a commonly used method for measuring material loss occurring in the interior of plant and pipelines. This technique operates by measuring the change in electrical resistance of a metallic element immersed into a flow stream or into a container of the medium. The net changes in the resistance ratio are solely attributable to metal loss of the exposed element. If the corrosion occurring in the vessel under study is roughly uniform, a change in resistance is proportional to an increment of corrosion.

The electrical resistance measurement monitoring system consists of an instrument connected to a probe. Probes are usually thin with small cross-sectional areas for increased sensitivity, and they are to be examined at fairly short intervals. Probes are usually positioned in such a way that allows replacement without interrupting the facility operation. The display instruments are calibrated in terms of metal loss, and reading is recorded on charts. The instrument may be permanently installed to provide continuous information, and it may be handy to gather periodic data from a number of locations.

Electrical resistance (ER) data are easily interpreted, and the technology is adequately supported by several commercial suppliers. Continuous corrosion monitoring and correlation with operational parameters are possible, provided sufficiently sensitive sensor elements are selected. ER probes are more suitable than coupons in the sense that results can be obtained without retrieval and weight-loss measurements. Although the combined thickness loss due to corrosion and erosion can be measured, the sensitivity of ER probes is insufficient to qualify for real-time corrosion measurements.

5.5 Linear Polarization Resistance

Polarization resistance is primarily used to quickly identify corrosion upsets and start corrective measures for the plant life extension. The technique is utilized to maximum effect when installed as a continuous monitoring system. This technique has been used successfully in the past in almost all types of water-based, corrosive environments, such as hydrocarbon production with wastewater treatment systems. LPR results are easily interpreted, and the technology is well supported by several commercial suppliers, including Rohrback Instruments. The diagram of LPR with two electrode polarization instruments is illustrated in figure 5.2.

Interpretation of the measurements is straightforward. Continuous online monitoring is possible because the measurements take only a few minutes. The high sensitivity of this technique facilitates real-time monitoring in appropriate environments. Polarization instruments work based on the principle that the amount of current required to cause the potential change is proportional to the corrosion current change, as presented in equation 5.1.

$$I_{crr} = K \left(\frac{I_{app}}{\Delta E} \right) \tag{5.1}$$

Here, I_{crr} is equal to the specimen corrosion current, ΔE is changes in electrical potential of the specimen, K is the collection of constant, and I_{app} is applied current required to change the electrical potential of the specimen. LPR measure the electrical potential of a specimen corroding in conductive fluid and current required to make the change, provided the potential change is not exceeded 10–20 millivolts. The technique is also based only on uniform corrosion principles. An environment with relatively high ionic conductivity is required for accurate measurements, but unstable corrosion potentials will produce erroneous results.

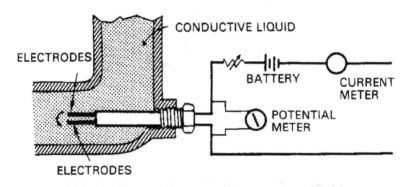

Figure 5.2 Two-electrode polarization instrument

5.6 Corrosion Potential Monitoring

Measurement of corrosion potential is a relatively simple concept, and it usually classified as an intrusive indirect method. It is broadly used for monitoring reinforcing steel corrosion in concrete structures. It is also used to monitor the effectiveness of cathodic protection as a means of the corrosion protection system, particularly the location of cathodic and anodic areas, indicating the direction and magnitude of current flow in an electrolyte. The diagram in figure 5.3 shows the cathodic protection potential measurement on a pipeline.

The corrosion potential is measured relative to a reference electrode, which is characterized by a stable half-cell potential. Therefore, a reference electrode has to be introduced into the corrosive medium for the measurements. Otherwise, an electrical connection has to be established to a structure in conjunction with an external reference electrode. The measurement technique and the required instrumentation are relatively simple. Although the technique may indicate changes in corrosion behavior over time, it does not provide any indication of corrosion rates.

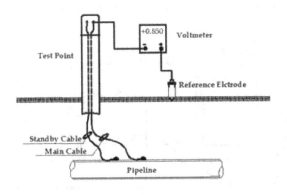

Figure 5.3 Corrosion potential measurements

5.7 Chemical Analyses

Various types of chemical analyses can provide valuable information in corrosion-monitoring programs. The measurement of pH, conductivity, dissolved oxygen, metallic and other ion concentrations, water alkalinity, concentration of suspended solids, inhibitor concentrations, and scaling indices fall within this area. Chemical analysis of producing oil or gas can sometimes determine the presence and amount of corroding agent, the corrosion product, and the nature and extent of corrosion. For instance, a test for iron may indicate corrosion activities.

The chemical analysis method makes it possible to perform corrosion monitoring cost-effectively and provides useful supplementary information to direct corrosion measurement techniques for identifying causes of corrosion damage and appropriate solutions. However, this technique may not provide direct information on corrosion rates and correlation with actual corrosion damage. Because this technique required laboratory measurement, the results may not be immediately available, and interference effects from other chemical species can also lead to inaccurate results.

5.8 Biological Corrosion Monitoring

Biological monitoring measures the presence of bacteria that consume sulfate and generate sulfuric acid (H_2SO_4) in a system. Sulfuric acid is known to be extremely corrosive to metal equipment. This is a common test often used to check for anaerobic bacteria corrosion caused by sulfate-reducing bacteria that produce hydrogen sulfide and form metal sulfides that corrode metals. When the corrosion product is treated with hydrochloric acid, it produces hydrogen sulfide, which can be detected by smell. This test should be immediately carried out on samples to avoid oxidation of sulfides when exposed to air.

5.9 Hydrogen Penetration Monitoring

Hydrogen penetration monitoring uses probes to measure the amount of hydrogen that dissolves into steel components. In the petroleum industry, hydrogen is often a by-product in many reactions and can lead to hydrogen-induced cracking or hydrogen embrittlement if left unattended. In acidic process environments, hydrogen is a by-product of the corrosion reaction. Hydrogen generated in such a reaction may be absorbed by steel, particularly when traces of sulfide or cyanide are present. This might lead to hydrogen-induced failure due to any mechanisms. The concept of hydrogen probes is to detect the amount of hydrogen permeating through the steel by mechanical or electrochemical measurement and to use this as a qualitative indication of corrosion rate. Hydrogen-monitoring sensors will be attached to the outside walls of vessels and piping with an instrument that instantly measures and records hydrogen reaction.

Hydrogen-monitoring probes are based on the following principles. Pressure increases with time in a controlled chamber, as hydrogen passes through the material and into the probe chamber or an electrochemical current resulting from the oxidation of hydrogen under an applied potential. Alternatively, the current may flow in an external circuit, based on a fuel cell principle, whereby hydrogen entering the miniature fuel cell causes the current to flow through the pipe radius being monitored, which creates a tight seal once it is welded onto the pipe. The saddle design offers a larger surface area in which gas may be captured. Measured output is related to the ambient hydrogen concentration, which can then be used to calculate an approximate rate of corrosion. The saddle hydrogen probe provides an instantaneous and nondestructive measurement of the hydrogen levels present in a system. The probe is maintenance-free due to the fact it has no moving parts, and its nonintrusive design makes it ideal for systems that undergo pigging.

The patch or saddle probe is affixed to the pipe that to be monitored. The hydrogen generated by an internal corrosion reaction will permeate through the pipe wall, be trapped by the patch body, and be detected by the sensing element. The instrument offers a high-integrity hydrogen probe that utilizes a nonintrusive design to measure the level of hydrogen present in a system. The probe is available in two configurations. The first is a patch unit, which is a probe body welded directly onto the pipe (fig. 5.4). The second option is a saddle assembly that consists of a saddle probe, which is welded onto the pipe (fig. 5.5).

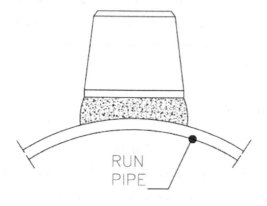

Figure 5.4 Patch probe

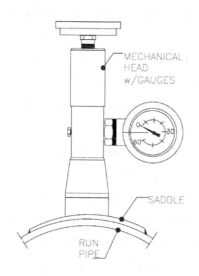

Figure 5.5 Saddle probe with mechanical head

Both styles are designed to capture hydrogen permeating through the pipe wall and channel this gas into the electronic or mechanical measuring head. Patch and saddle probes are typically manufactured from material similar to the piping being monitored. The patch and saddle assemblies are designed to be welded onto the pipe despite the fact that the hydrogen patch probe can be installed to measure hydrogen produced by corrosion of vessels and piping. Nevertheless, the reading is not relatively an indicator of the steel material conditions because not all the hydrogen produced by the system enters the steel structures.

5.10 Visual Inspection

Visual inspection is the oldest and most widespread form of corrosion inspection. The visual inspection principle is based on the visible reflection of light from the surface to expose some of its facial appearance, particularly corrosion damage conditions. Several corrosion forms like uniform corrosion, exfoliation, pitting, and intergranular corrosion can be exposed by visual inspection when appropriate access to the area is possible. Apparently, visual inspection can detect only surface anomalies. However, some internal corrosion processes do produce surface indications like pillowing or flaking of a metal surface. Visual inspection is a simple and cheap method of identifying corrosion defects before the failure occurred, depending upon the experience of the personnel. Visual inspection can be carried out either directly or remotely.

The remote method can be done using bore scopes, fiberscopes, or video cameras. Visual inspection is also frequently conducted using a strong flashlight, a mirror, and magnifying aids. The magnifying glass is particularly recommended for identification of suspected cracks and corrosion. The disadvantage of visual inspection is that the surface must be fairly clean and accessible to either the naked eye or optical aids. Another big disadvantage of visual technique is that it is labor-intensive and does not provide quantitative assessments of material loss.

5.11 Radiographic Examinations

Radiographic examination is a nondestructive technique of inspecting materials for surface and subsurface discontinuities. In a circumstance that a structure surface is not practically accessible for examination by other available monitoring means, radiographs using X-rays or radioactive isotopes can be used to evaluate metal corrosion, both of which are electromagnetic waves with very short wavelengths.

The waves penetrate the material and are absorbed, depending on the thickness or the density of the material being examined. By recording the differences in absorption of the transmitted waves, variations in the material can be detected. The variations in transmitting waves may be recorded by film or electronic devices, providing a two-dimensional image that requires interpretation. The method is sensitive to any discontinuities that affect the absorption characteristics of the material.

Radiographic pictures provide accurate information on metal conditions, and they can be applied to confirm the results of other tests using different monitoring methods. Advances in the use of radiography are being made that involve using computers and high-powered algorithms to manipulate the data. The advantage of this method is that it provides pictures that can be understood by technical and nontechnical personnel.

5.12 Electromagnetic Examination

The method is used to examine ferromagnetic wire rope products in which the magnetic flux and magnetic flux leakage methods are used. The magnetic flux method is capable of detecting the presence, location, and magnitude of the metal loss from wear, broken wires, and corrosion. The magnetic flux leakage method is capable of detecting the presence and location of flaws such as broken wires and corrosion pits. An electromagnetic inspection tool also identifies damages in structural members, such as well casing. The device shows damages both inside and outside of the casing. The electromagnetic examination device makes up four channels. Two out of these four channels are for flux leakages. The third channel provides records of an abnormality inside the casing, and the fourth channel indicates whether the abnormality is circular.

Understanding the electromagnetic logs requires specific information and detail on the well, casing, and tubing specifications. Accurate interpretation of the logs requires the precision of well data and a detailed record of the well's construction.

5.13 Calipers Survey

A caliper survey is used for surveying the insides of casing and tubing for mechanical damages. Multifingered calipers are well-established tools for evaluating internal problems of tubing, but they provide no data about external flaws, and the methods are affected by scale buildup on the inner wall. Calipers are fitted with feelers, which are held by spring pressure to tubing such as casing walls. The number of feelers depends on the pipe size. Tubing calipers may have feelers range from twelve to forty, and casing calipers may range from forty to sixty.

The feelers extended into the corrosion pit or rod score, mechanically change the setting of the potentiometer above the feeler assembly, and record the deflection on a metal chart. The data are transmitted through a conductor to an instrument on the surface. Caliper tools also utilize a set of mechanical fingers that ride against the internal surface of a pipe under survey. Caliper pigs are used to measure pipeline internal geometry. They have an array of levers mounted on the body of the pig. The levers are connected to a recording device in the pig body. The body is normally compact, about 60 percent of the pipe's internal diameter, which, combined with flexible cups that allow the pig to pass constrictions up to 30 percent of the bore. The limitation of caliper survey is that scale or hard corrosion products may fill pits and prevent the feelers from entering and accurately measuring.

5.14 Pulse-Echo Technique (Ultrasonic Devices)

Ultrasonic nondestructive testing, known as ultrasonic NDT, is a method of characterizing the thickness or internal structure of a test piece through the use of high-frequency sound waves. Ultrasonic devices provide recorded information on metal thickness, surface condition, and flaw detection. This technique is extremely sensitive and capable of scanning over a large surface area using ultrasonic waves. The method is predominantly used to conduct surveillance and integrity surveys on oil and gas installations to ascertain the degree of corrosion damage on structural member and vessels.

The ultrasonic device uses ultrasonic energy generated by transducers, which change high-energy frequency signals into high-frequency mechanical energy. A liquid couplant to the metal wall transmits the sound wave generated by a transducer, and ultrasonic sounds travel through the wall until they reach the discontinuity, which is the end of the testing specimen. The echo sound waves are received and transformed into electrical impulses by the transducer. The device measures the time between the impulse and its reflection, and hence the thickness of the metal can be calibrated, read, and recorded on data sheets.

The setup of analog ultrasonic equipment can be seen in figure 5.6. The pulse generator sends an electric pulse to the transmitter probe, which produces an ultrasonic pulse. This ultrasonic wave spreads into the specimen and is reflected to the receiver, which transforms the wave into an electric signal. This signal is then sent to the amplifier and from there to the cathode ray (CR) tube, which displays the signal as peaks.

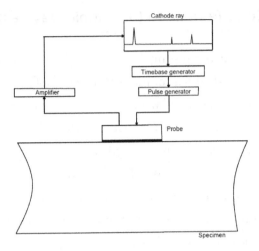

Figure 5.6 Setup of Analog Ultrasonic Equipment

The horizontal axis is proportional to the time t. The vertical axis shows the amplitude of the signal. The time base generator produces a high-frequency wave and makes the spot move across the CR tube. The pulse-echo method has the advantage of being a single contact method, requiring access from one side only, but it has the disadvantage of requiring samples of lower thickness compared with the through-transmission technique. Figure 5.7 shows OmniScan iX NDT Ultrasonic equipment from Olympus NDT Incorporation.

Figure 5.7 OmniScan iX (Olympus NDT Incorporation)

CHAPTER 6

CORROSION CONTROL METHODS

6.0 Introduction

Corrosion control is simply the prevention of deterioration of material caused by its environment. This control can be carried out by changing the material environment, altering the material composition (alloy), or placing a barrier between the material and its environment. The methods of corrosion control are variations of these general measures. However, corrosion engineers often combine more than one of these methods for effective corrosion protection.

Corrosion control and safety of engineering facilities go hand in hand. For instance, experience shows that corrosion damage is identified to be the primary cause of pipeline and related engineering installation accidents worldwide. The result of pipeline failure can be destructive to life and property, leading to pollution of the environment. To prevent such catastrophe, the operators of facilities that are prone to corrosion damage must have been using corrosion control as one of their strongest weapons in the encounter against corrosion failures.

In nearly all circumstances, metal corrosion can be managed, slowed, or even stopped by using the proper techniques. Corrosion prevention can take a number of forms depending on the circumstances of the metal being corroded. Corrosion prevention techniques can be generally classified into six groups, as presented in table 6.1. The techniques provide cost-effective solutions to corrosion problems by selection of appropriate corrosion control options. No metal is immune to corrosion in all environments, but understanding the environmental conditions that are the cause of corrosion, and changes in the type of metal being used, can also lead to significant reductions in corrosion. This single act may save billions of dollars in infrastructure repair and extend useful service life.

Table 6.1 Corrosion control options

S/N	Options	Applications Descriptions
1	Design	Use codes/specifications during engineering, assess corrosion during concept and detail design (shape, compatibility and surface condition), and introduce QA/QC and inspection procedures
2	Environmental Control	Modify moisture/humidity levels, change pH, and lower oxygen concentrations.
3	Material Selections	**Material Examples** Carbon-manganese steels CRAs (stainless steel, nickel alloys), clad steels **Selection factors:** Strength, weight, ease of fabrication and construction, availability, durability, and relative costs

S/N	Options	Applications Descriptions
4	Coatings and Painting	**External** Marine and atmospheric **Internal** Immersed and acid resistance **Major Factors** Cost, availability, ease of surface preparations and application, inspection, life expectancy, resistance to degradation and high temperature, repair options, and maintainability
3	Chemical Treatments	Inhibition for oil/gas production systems, use of biocides in water systems, hydro test and annular fluids, scale prevention in flow lines, and cooling systems
4	Cathodic Protection	**Method** Electrochemical means of corrosion control Sacrificial and impressed current systems **Application** Immersed, buried structures and pipelines
6	Process Control	Change throughput, flow rate, temperature

6.1 Design and Environmental Control

Corrosion control begins with system design. The application of rational design principles can eliminate many corrosion problems and greatly reduce the time and cost associated with corrosion damage. The properties of produced fluids and the rates at which they flow through piping are important considerations in planning a production system. The flow rate may be a difficult factor to evaluate in the design stage. Rapid, turbulent flow of oil in pipes can prevent brine damage, keeping the brine dispersed and preventing it from separating out. In such cases, the pipe should be sized such that oil flow in practice is turbulent and continuous. When corrosive oil is to be handled, pipes should be arranged to avoid dead ends where scales and brine can collect and remain undisturbed. The following steps may be taken as an environmental control and modification in oil and gas facilities to combat corrosion damages.

- Drying and mechanical removal of water
- Removal of corrosive agents such as H_2S and CO_2
- Temperature control such that heating and cooling is above the dew point
- Filtration of sand and other debris

Corrosion often occurs in dead spaces or crevices where the corrosive medium becomes more corrosive. Such areas can be eliminated in the design process. For the structural component, where stress-corrosion cracking is possible, the component can be designed to operate at stress levels below the threshold stress for cracking. Also, for the equipment components where corrosion is anticipated, such components may be standardized and designed to provide for higher interchangeability. In addition, in case of critical items like pumps, extra equipment may be installed to allow for the maintenance of one unit while the other is operating.

6.2 Material Selection

Based on a technical point of view, an obvious prevention to corrosion problems should be the use of more resistant materials. In many cases, this approach is an economical alternative to other corrosion control methods. The selection of a particular material is the outcome of several considerations and agreements during engineering design phases. The choice of a material is typically as a result of several compromises, and the final selection will be in an accord between technical competence and economic factors.

MATTHEW OMOTOSO, PH.D

Each metal and alloy have unique and inherent corrosion behavior that can range from the high resistance of noble metals to the low corrosion resistance of active metals. However, the corrosion resistance of metal is strongly dependent on the environment corrosivity to which the metal is exposed, such as temperature and chemical composition. The classic relationship between the rate of corrosion, the corrosivity of an environment, and the corrosion resistance of a particular material is represented by equation 6.1.

$$R = \frac{E_{corr}}{M_R}$$
(6.1)

Here, R is the rate of corrosion attack, E_{corr} is the corrosivity of environment, and M_R is the corrosion resistance of the metal. For a given corrosion resistance of a material, as the corrosivity of an environment increases, the rate of corrosion increases. Also, as the corrosion resistance of metal increases, the rate of corrosion decreases. In practice, an acceptable rate of corrosion rate of corrosion is fixed, and the question is to match the corrosion resistance of the material and corrosivity of the environment to be at or below the specific corrosion rate.

Oftentimes, there are several competing materials that can meet the corrosion requirements, but the material selection process becomes one of the determining factors. In this aspect, there is a need to identify which of the candidate materials provides the most economical solution for a particular service. In recommending a material for any engineering application, the material needs to be defined and evaluated, and at the end the client chooses the most economical one. The material selection process also influences whether it will be used for a new construction or for repairs of an existing facility. With regard to a facility repair purpose, there might be less opportunity for the use and selection of a new material. The major considerations will be centered on simplicity of the fabrication, installation method on the site, and the residual life span of the facilities to avoid overdesigning. In case of a material for a new construction, the selection process needs to be started early before the design is finalized by considering the following listed factors, which are well represented and guided by industrial standards, such as NACE, API, ASME, ASTM, NORSOK, and DIN.

- Mechanical properties of material (strength, ductility, etc.)
- Corrosion resistance bearing of material
- Service temperature and chemical resistance
- Fabricability and welding property of material
- Material thermal conductivity
- Material availability and cost

Construction material is known to have mechanical properties codes, but there are no such codes governing corrosion properties, unless for the recommended practices published by NACE.

Nevertheless, the extra cost associated with the selection of improved corrosion resistance material is invariably smaller than the high maintenance costs and premature failure of the facilities. In the absence of adequate corrosion resistance material and provision of corrosion, allowance components may not meet the expected design life span before failure occurs.

6.3 Coating System

The coating for corrosion protection can be divided into two broad groups as metallic and nonmetallic or organic and inorganic. The fundamental function of either metallic or nonmetallic coating is to isolate the underlying metal from the corrosive media. Individual coatings are formulated to perform specific functions and must be selected to become component parts of the total system, designed for optimum results considering the environment and service expectations.

The primary function of organic coatings in corrosion protection is to isolate the metal from the corrosive environment and forming a barrier layer to suppress corrosion process. Inorganic coatings include porcelain enamels, chemical-setting silicate cement linings, glass coatings, and zinc and phosphate. But the most effective is zinc silicate primers that become anodic to the steel in a corrosion cycle. The advantage of inorganic coating is that it disallows rust creep and coatings undercutting around the damaged areas. Inorganic coating such as carbides is used for wear-resistant application, and silicide are also used for heat-resistant purpose.

There are different types of anticorrosion-pigmented primers available in inorganic coating category that passivate the steel. A typical coating system may include a primer, an intermediate coat, and a top coat. The methods of applying an inorganic coating are plating, thermal spraying, nitriding, and carburizing. Inorganic coatings are produced by chemical action, with or without electrical assistance. The treatments change the immediate surface layer of metal into a film of metallic oxide or compound, after which the metal has better corrosion resistance than the natural oxide film, which provides effective base for supplementary protection such as paints.

Organic coatings are coatings made from organic structure having epoxide or oxidase endcaps. Examples of such endcaps are epoxies, urethane, and polyurethane. They are either air-dried or heat cured. Epoxy coatings generally provide protection to substrates by forming a barrier to the environment and keeping the electrolytes necessary for corrosion at bay. Minimum film thickness is normally obtained in one or two coats. Two coats are typical to offshore application to ensure a high level of protection and a minimum amount of film defects.

Epoxies are generally more abrasive and chemical resistant, and they protect not only the substrate itself but also the zinc primer from detrimental factors. Although epoxies are widely used in direct-to-metal applications, when combined with zinc rich primer, it can provide the best possible long-life, anticorrosive protection in an atmospheric exposure. However, the disadvantages of epoxy coatings are poor resistance to ultraviolet from sunlight, quickly chalking, and fading rapidly. This leads to erosion of the coating's film thickness and reduction of the coating effectiveness.

Top coats are generally required to have high resistance to ultraviolet from sunlight, in addition to its resistance to other adverse environmental factors. Polyurethane coatings are generally acknowledged as providing optimum resistance to ultraviolet with high degrees of flexibility and chemical resistance. They also help to maintain a very high level of cosmetic gloss and color retention, and they can be cleaned very easily, generally with 7–9 pH detergents and freshwater washing. Although polyurethane finishes offer no real anticorrosive or barrier protection to the substrate, they do provide high level of protection to the integrity of the coatings system.

Coating application for industrial structures is defined by its protective, rather than its aesthetic, properties. The most common use of industrial coatings is for corrosion control of steel structures because it is applicable to offshore platforms, bridges, and underground pipelines. Additional functions of coating are metallurgical surface modification (hardening), wear resistance, and intumescent coatings for fire resistance. The most common polymers used in industrial coatings are polyurethane, epoxy, and moisture-cure urethane, as shown in table 6.2.

Table 6.2 Application of polymers coating systems

Type	Area of Use	Temperature Limit (°C)
Asphalt	Onshore and Offshore	-10–70
Coal Tar	Onshore and Offshore	-10–70

Type	Area of Use	Temperature Limit (°C)
Fusion Bonded Epoxy	Onshore and Offshore	-10–70
Polyethylene	Onshore	-40–80
2 Layers of Polyethylene	Onshore	-40–80
3 Layers of Polyethylene	Onshore and Offshore	-40–120
3 Layers of Polypropylene	Onshore and Offshore	-40–120
Polyurethane Tar	Bends and Fittings	-20–70
Rubber Linings	Splash and Tidal Zone	-40–120
Metal Sprayed (Al)	Very Severe Atmosphere	-60–450

Engineering facilities such as crude oil production platforms, pipelines, and bridges must be designed to protect the essential and hyper expensive assets from corrosive environments. The costs of protecting capital infrastructure should not be compromised by disregarding protective coatings. The durability of engineering facilities is a challenge to not only the operator but also the coating manufacturers to protect the assets from ambient conditions, which are destructive to coating systems. Lack of routine maintenance mandates that coating systems should be engineered for high-level performance and maximum service life. The coating should be designed to enhance personal safety, visibility color markings, and fire and heat resistance.

Different coatings may be suitable for different environment and facility locations. There are several acceptable coatings in the industry, and their method of application depends primarily on the operating temperature and the cost. In selecting coating for underground engineering facilities (pipeline, etc.) application, due consideration should be given to the following factors.

- The soil's resistivity
- Level of groundwater table throughout the year
- For cohesive clay soil, data on pipeline-to-soil friction
- In rocky areas, consideration for damage to the coating
- In tropical locations, termite attack
- Potential damage to plant-applied coating in transit to job site
- Additional coating thickness at highway and railroad crossings
- Service life anticipated for the pipeline
- Comparative coating quality over the pipeline service

The splash zones of marine structures are subjected to extreme corrosion damages due to constant wet/dry conditions and abrasion from service boat unloading crews and supplies. Thus, special coatings are designed to form a highly abrasive-resistant barrier and in many cases incorporate the use of glass beads, quartz, aluminum flakes, and other inert pigmentation to enhance impermeability and abrasion resistance. These barrier coatings are applied in ultra-high film thickness direct to metal without any top coat, but they require specialized application equipment in many cases. They are also formulated to cure very quickly, be compatible with damp or wet conditions, and in some cases cure underwater. This allows maintenance application in an offshore environment in the presence of sea wave action.

The maintenance of safe working conditions is paramount on marine facilities such as oil and gas production platforms as a result of harsh environment and risky nature of work. Therefore, specialized coatings are to be provided in the walkways, exterior work areas, and boat landing platforms. The recommended specialized coating should be durable and chemically resistant to fuels and hydraulic fluids, and have a high degree of slip coefficient. Coating specifically designed with anti-slip properties normally incorporates with coarse aggregates. Alternate materials are also available for anti-slip deck coverings that may be applied over old and intact traditional coatings. The Epoktreadtile system is one method that can be applied easily and quickly, with the deck being returned to service immediately after application. Advantages of coating application include reduction of abrasive, adhesive, and crater wear on the tools for conventional wet, dry, and high-speed machining parts. Coating technology ensures an increase in the facilities' life span through reduction in corrosion and abrasion damages.

6.4 Pipeline External Anti-Corrosion Coating

The external coating of a petroleum product consists of a layer of epoxy base primer (fusion bonded epoxy), a layer of polymeric base adhesive, and an external multiple layer of polyethylene (PE) compound, as illustrated in figure 6.1.

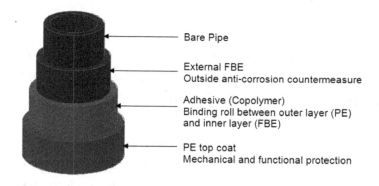

Bare Pipe

External FBE
Outside anti-corrosion countermeasure

Adhesive (Copolymer)
Binding roll between outer layer (PE)
and inner layer (FBE)

PE top coat
Mechanical and functional protection

Figure 6.1 Illustration of the three coating layers

The minimum thickness of epoxy primer, polymeric adhesive, and polyethylene are 0.15–0.25 mm, 0.15–0.25 mm, and 0.20–0.30 mm, respectively. The polyethylene thickness varies depending on the diameter and the total coating thickness, which shall not be less than 25 mm. The primer not only has the function of ensuring good anchorage of the overlying coating fraction to the pipe metal, but most important, it reduces the problems deriving from excessively high negative polarization of the metal. The first coat of coating layers on metallic pipeline is primer, and its functions are not limited to the following.

- Good similarity and compatibility with the polymeric adhesive it can react with it, such that the reciprocal anchorage is increased
- Primer should have low viscosity and high metal wettability at its application temperature, as well as have high adherence to the metal itself and to the polymeric adhesive layer
- By nature and formulation, primer should be practically water impermeable, have extremely low water absorption, have water vapor transmission coefficients, and be resistant to the attack of microorganisms present in the application medium
- Primer should be suitably stabilized against aging at the maximum working temperature
- Primer should consist of a suitable fusion bonded epoxy (FBE) resin base

The polymeric adhesive is the second layer of coating on the pipeline, and the function is ensuring strong anchorage for the protection layer to the primer coat. However, other functions of polymeric adhesive include the following.

- Possess good affinity and compatibility with the primer and the protection layer so that the reciprocal anchorage is increased
- Have high adherence to both layers; possess strong cohesion in the specified temperature range
- Be stabilized to heating to maximum extrusion temperature and at continuous maximum working temperature
- Be free of solvents or substances that are volatile or unstable at the specified temperatures
- Consist of polymers having good affinity and compatibility with the protection coat mix

The top coat should consist of at least 97 percent in mass of ethylene homopolymers or copolymers, or it should be a mixture of both with mix density of about 0.94 g/cm^3. The top coat is black in color and contains around 2–3 percent in mass of carbon black, uniformly and finely distributed with one or more additives. This is to ensure stabilization against oxidation and continuous working temperatures to which it could be exposed. The melt flow index, as specified in ISO 1133 (2005), shall never exceed 0.25 grams per 10 minutes. The carbon black must have an iodine number exceeding 110, as recommended in ISO 1304 (2016). The additives must be compatible with the polymers in which they are dispersed to prevent them from coming to the surface of the coating, and they should not cause significant mass loss due to prolonged exposure to heat conditions.

Before the application of coating, pipeline shall be free of wrinkles and generally smooth, with no air bubbles. The coating materials shall be applied strictly in accordance with the manufacturer's recommendations. Epoxy powder application occurs through fusion by means of guns to spray the material on the preheated pipe surface. Polymeric adhesive and polyethylene will be applied by side-band extrusion, with the overlapping of more layers, or by circular extrusion in one layer only. The application of the coating at any point of the pipe shall never be interrupted. Should this occur, it will not be allowed to restart the application, and the pipe shall be grit-blasted again. The pipe surface shall be free of any temporary protective paint or markings that cannot be removed during normal cleaning operations. The pipe shall be correctly identified throughout the coating process, making sure that they are not damaged in any way. Damaged pipes shall be quarantined and immediately identifies to determine their acceptability; otherwise, they are rejected.

Pipes shall be thoroughly cleaned of grease, oil, and other foreign matter. For a heavily contaminated environment, it may be necessary to use a wire brush or power water washing of about 203 kgf/m^2. All water used shall be potable quality and demonstrate that there is no salt contamination on the pipe surface prior to blasting. The steel surface may be preheated to ensure that all traces of moisture are eliminated, and the chemical analysis of the abrasion should be verified that it is free from carbonates and chlorides. The surface profile of the steel surface after abrasive blast cleaning shall be measured and retained with the inspection records, by means of the replica tape method or an equivalent, and the surface profile shall be in the range of 50–100 microns.

Any surface defects highlighted by the blast cleaning, such as scales, laminations, burrs, and gauges, shall be removed by grinding with an abrasive disk. If the repaired area exceeds 25 cm², then the area shall be spot-blasted to restore the surface profile. All defective pipes shall be stored in a separate area. For the badly defected areas repaired by grinding, the parent metal shall be assessed for remaining wall thickness to ensure that the pipe is still within the specification limitations for service. In case of any surface rusting during the repair process, the pipes shall be returned for abrasive blast cleaning.

Every coated pipe shall be 100 percent visually inspected. The coating shall be black in color, uniform, and homogeneous. The pipe shall not show any defects such as wrinkles, indentations, cuts, swellings or material thickening, air blisters, and tears. Any defects found during visual inspection shall be repaired. All coated pipes shall be subjected to 100 percent holiday detection, to be carried out when the coated pipe temperature is below 90°C using a high-voltage DC impulse-type holiday detector provided with an audible signal. The performance of the detector shall be checked using a pipe that is artificially pinholed, and the permitted coating holidays shall not be in excess of two on a pipe joint. The final thickness measurements shall be carried out on at least 10 percent of the coated pipes in every shift. The minimum and maximum detected values shall be recorded. The thickness shall be measured on at least twelve points uniformly spaced over the length of the pipe and the circumference using electromagnetic equipment. The minimum detected values shall not be lower than the minimum specified value.

6.5 Galvanizing

The word *galvanization* was used for the management of electric shocks in the nineteenth century. It started from Galvani's induction of jerks in severed frogs' legs by his accidental generation of electricity. There was a belief that electric shock is beneficial to health. But this fact is basically refuted apart from some limited uses in psychiatry, such as electroconvulsive therapy. Afterward, the word *galvanic* was used for processes of electrode position, which is still widely applied technology today. However, the term *galvanization* has found to be mostly associated with zinc coatings to protect other metals from corrosion damages. Galvanic paint, a forerunner to hot-dip galvanization, was patented by Stanislas Sorel in France in 1837, and the first acknowledged case of galvanizing of iron was found in the seventeenth century in the Royal Armouries Museum collection in India.

Galvanizing is the process of applying a zinc coating to fabricated iron or steel material by immersing the material in a bath consisting primarily of molten zinc. Although galvanization can be done with electrochemical and electrodeposition processes, the most common method in current use is hot-dip galvanization, in which steel parts are submerged in a bath of molten zinc.

The term *galvanizing* is also known as application of zinc coating by the use of a galvanic cell, which is also referred to as electroplating. The practical difference between the hot dip and electroplating is that hot-dip galvanization produces a thick, durable, and matte gray coating, whereas electroplated coatings tend to be thin and brightly reflective. As a result of zinc lack of thickness in electroplated coatings, it depletes faster and makes the method less suitable for open-air applications. However, when electroplating application is combined with painting that may slow down zinc depletion, the method may be durable enough for outdoor application. Electroplating is frequently used in many outdoor applications because it is cheaper than hot-dip zinc coating and appears fine when newly applied.

The easy method of galvanic application process is a distinct advantage over other methods of providing corrosion protection. The significance of galvanizing is as a result of the corrosion resistance of zinc, which is considerably greater than that of steel. For galvanized steel, the zinc serves as a sacrificial anode and cathodically protects exposed steel. When the coating (zinc) is scratched off, the exposed steel will be protected by the leftover zinc that stands as sacrificial anode.

This phenomenon is a great advantage over other methods such as paint, enamel, and powder coating. The galvanizing method of material protection is better than others because the technique is inexpensive, is simple to apply, and has a relatively long, maintenance-free service life.

6.6 Anodizing

Colonialanodizing.com and Wikipedia state that anodizing was first used on an industrial scale in 1923 to protect Duralumin seaplane parts from corrosion. The first sulfuric acid anodizing process was patented by Gower and O'Brien in 1927.

Anodizing is an electrolytic passivation process used to increase the thickness of the natural oxide layer on the surface of metal. The method entails electrolytic oxidation of a surface to produce a tightly adherent oxide scale, which is thicker than the naturally occurring film. The difference between electroplating and anodizing is that the oxide coating is integral with the metal substrate as opposed to being a metallic coating deposition.

The oxidized surface is hard and abrasion resistant. It provides some degree of corrosion resistance and better adhesion for paint primers and glues than bare metals. An anodic film is also use for cosmetic effects like transparent coatings that are capable of effecting light. Anodizing is also used to avert galling of threaded parts and making of dielectric films for electrolytic capacitors. Anodic films are mostly useful to protect aluminum alloys; however, processes also exist for titanium, zinc, magnesium, niobium, zirconium, hafnium, and tantalum.

The anodizing process changes the surface microscopic texture of the crystal structure of the metal near the surface. For instance, the anodized aluminum surfaces are harder than original aluminum but have moderate wear resistance that can be improved with increasing thickness. The anodic films are generally much stronger and more adherent than most paint and metal electroplating, but they are more brittle, which makes it less likely to crack and peel off due to aging and wear.

The processes of anodization of wrought alloys include cleaning in either a hot soak cleaner or in a solvent bath due to the presence of intermetallic substances. The anodized aluminum layer is grown by passing a direct current through an electrolytic solution, with the aluminum object serving as the anode. The current releases hydrogen at the cathode and oxygen at the surface of the aluminum anode, creating a buildup of aluminum oxide. The voltage required by various solutions varies, and higher voltages are typically required for thicker coatings. Special conditions such as electrolyte concentration, acidity, solution temperature, and current should be controlled to allow the formation of a consistent oxide layer. Harder, thicker films tend to be produced by more dilute solutions at lower temperatures with higher voltages and currents.

6.7 Inhibitors

Inhibitors can be defined as the substances that slow or prevent chemical reactions. For example, a corrosion inhibitor is a chemical compound that decreases the corrosion rates of a material, typically a metal or an alloy. A common mechanism for inhibiting corrosion involves formation of a passivation layer that prevents access of the corrosive substance to the metal. However, permanent coating such as plating is not usually considered for inhibitors because corrosion inhibitors are additives to the fluids that surround the metal to be protected. Basically, the nature of corrosive inhibitor depends on the material being protected, which is usually a metal object, and the corrosive agents to be neutralized. Inhibitors also depend upon their molecular structures for effectiveness. The cations or anions are joined to form a massive organic ion that is chemically inactive. The active parts of these molecules are attracted to either the cathodic or the anodic metal surfaces. The monomolecular film is a strongly bonded to the metal by a process called chemisorption and, the film can attract oil that is thick enough to be visible. Oily films have high electrical resistance, and they separate metal surfaces from electrolytes and cut short electrical circuits.

High temperatures, rapid flow of well fluid, brine, sand, and rubbing action of sucker rods and wire lines can weaken the thin film inside the oil well. Most oil field inhibitors are liquids and are soluble in either oil or water. Inhibitors are used in oil industry such as crude oil wells, where the quantities of the agent applied are not sufficient to neutralize the corrosion agents. Therefore, the durability of inhibitor films is an important quality during selection stages.

Glycol or methanol treatment is considered for corrosion control in wet gas pipelines when it is added to the wet gas, typically at a pipeline inlet. The chemical absorbs water from the gas phase and reduces the dew point temperature. Consequently, the relative humidity of the gas is also reduced. Methanol is expensive but very effective and preferred at cryogenic conditions. Ethylene glycol is less expensive and easily recovered except in very low temperatures, where its viscosity is very high. It is less soluble in the liquid hydrocarbons that are common in producing field gas system.

6.8 Pipeline Pigging

A pig, also called a scraper or go-devil, has a vital part in the maintenance of pipelines. A pig is like a free piston that is moved with the oil through the pipeline and sealing, aligned with the inside wall and with a number of sealing elements. The moving force is due to the stream of oil in the line. Pigs execute a number of tasks such as cleaning debris from the line, removing residual product, and removing accumulated water, which is one of corrosion's agents.

Pigs required specially designed receiving and launching vessels, as shown in figures 6.2 and 6.3, respectively, to initiate them into the pipeline. These vessels consist of a quick opening closure for access, an oversized barrel, a reducer, and a neck pipe for connection to the pipeline. Pigs can be located using fixed signalers along the pipe or electronic tracking systems mounted inside the pig.

The pigging of pipeline will save the existing installation through positive internal corrosion prevention. There are additional benefits, which include the following.

- Prevention of pipeline blockage and corrosion attack
- Guaranteed maximum throughput
- Increased pipeline flow
- Reduce maintenance
- Clean product delivery
- Increased pipeline life
- Rapid return on investment

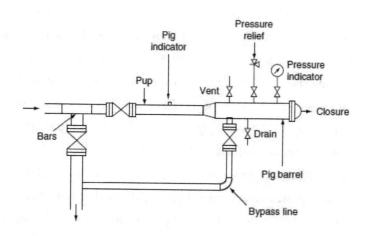

Figure 6.2 Pig receiver nozzle details

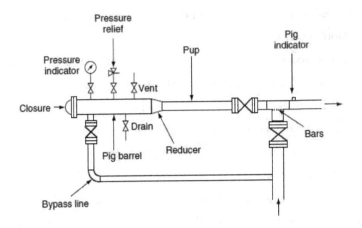

Figure 6.3 Pig launcher nozzle details

The operator must consider a number of criteria when selecting the proper pig for a pipeline: defining what task the pig will be performing, the pig's size, and the operating conditions. The pipeline layout is also important when choosing a pig because every pipeline has different configurations. There is no set schedule for pipeline pigging, but the quantity of debris gathered in a pipeline, along with the amount of wear and tear, can increase the frequency of pigging. Therefore, today pipeline pigging is carried out during all phases of the life of a pipeline.

Application of pipeline pigging can be for different purposes. It can be used simply to clear the line and to remove unwanted materials, such as wax and debris from pipeline. Pigs can also be used for inline inspection and examining the pipeline from the inside. Specialty pigs are used to plug the line or isolate certain areas of pipelines. Additionally, sealing pigs are used to remove liquids from the pipeline, and they can serve as an interface between two different products within a pipeline.

There are different types of utility pigs, which include mael pigs, foam pigs, solid cast pigs, and spherical pigs. Pipelines that have been coated internally cannot be cleaned with brush pigs because this would potentially damage the coating. In this case, the only method by which debris can be

MATTHEW OMOTOSO, PH.D

removed is by using pig discs and water flow. By the same token, however, coated pipelines tend to have less debris to clear out because rust and scale in particular should not be present in a significant way.

Pigs cover a range of pipeline operation functions. Foam swabs are used for drying operations, brush pigs may be used for cleaning, and standard sealing pigs are used for batching and separation. Gauge pigs are used as a simple inspection technique, and more sophisticated electronic caliper pigs may be used for a similar purpose. Intelligent pigs may be used to provide baseline data.

6.9 Corrosion Control in Reinforcement Concrete

Reinforced concrete is a composite material and is made up of concrete and reinforcement. Concrete is a mixture of cement, aggregates and water to form a relatively low tensile strength and ductile material. The embedded reinforcement has higher tensile strength and has good ductility. Reinforcing schemes are generally designed to resist tensile stresses in the regions of concrete that might cause unacceptable cracking and eventually structural failure. Concrete and reinforcement work together effectively as a composite material due to many reasons. The concrete coefficient of thermal expansion is similar to that of steel, thereby eliminating large internal stresses as a result of differences in thermal expansion and contraction. The hardening of cement paste within the concrete conforms to the steel surface characteristic, thus permitting stress to be efficiently transmitted between the two materials. Also, the alkaline chemical environment provided by the hardened cement paste provided passivation film on the steel surface, thereby making the steel more resistant to corrosion than it would be in other conditions.

Reinforced concrete structures are designed and built for a particular service life, as defined by the characteristic of the structure and the design aim of the engineer. For this period of time, the reinforced concrete structures are not expected to deteriorate to the extent that it cannot fulfill its primary functions. Basically, the deterioration of concrete structures is largely due to reinforcement corrosion damage when they are in contact with a high amount of aggressive corrosion agents. The reinforced concrete damage is especially large in the structures exposed to marine environment, contaminated groundwater, or deicing chemicals.

One of the common approaches employed to mitigate corrosion of steel in concrete is to provide adequate concrete cover in the threshold thickness of 50–75 millimeters, depending on the environmental exposure zone. Concrete corrosion can also be controlled by a low water-cement ratio that slows down capillary penetration of seawater and consequently minimizes penetration of chloride by ionic exchange. The addition of water-reducing agents and of minerals such as fly ash and silica fume proved to be effective for delaying initiation of corrosion and prolonging the useful life span of marine structures.

Prevention of reinforcement corrosion is partly achieved by recommending appropriate thickness concrete cover. The traditional design of concrete with regard to cement dosage and water-cement content ratio depends on the expected strength and the environment to which the concrete is exposed. Also, the thickness of concrete cover depends on the aggressiveness of the environment. Besides the other requirements given by the engineer, it is important to pay attention to reinforcement positioning, concrete mixing, and curing. The first line of defense against the reinforcement corrosion is high quality concrete. Nevertheless, high-quality concrete may crack after a while, leading to chloride penetration, even with low-permeability concrete. Thus, the final line of defense against reinforcement corrosion is that reinforcement itself being protected against contact with corrosion agents.

Since the time corrosion of reinforcement was known as a problem, many methods have been developed to prevent it from occurring. These processes can be grouped according to the method and procedure. Some of the common methods are the enhancement of reinforcement concrete structures designed, using materials that electrically isolate the steel from the concrete. The use of material with higher corrosion thresholds than conventional steel is now introduced to the reinforcement concrete design. Research works have been carried out at the University of Lagos, Nigeria, using bamboo as an internal reinforcement of concrete columns; Salau et al. (2012) has given credence to this assertion.

Also, a number of studies have been undertaken for the possibility of using fiber-reinforced plastic as a noncorroding reinforcement option for concrete. More than fifteen highway and pedestrian bridges have been constructed worldwide using fiber-reinforced plastic bars. Barrier techniques are proposed as other ways of protecting reinforced concrete from corroding by preventing water and chloride ions from penetrating the concrete, thereby slowing the corrosion process. In this facet, applications of good quality, low water/cement (w/c) ratio concrete, and adequate cover are to be used as a standard methodology in reinforced concrete structures construction. The provision of waterproof membranes with asphalt overlays on reinforced concrete building roof slabs and bridge structures are popularly used to prevent water from penetrating the concrete.

Electrochemical methods are also used for the protection of reinforcement in concrete against corrosion. Such methods include cathodic protection and electrochemical chloride extraction. These methods have the ability to stop corrosion in chloride-contaminated concrete, which has been used in several areas including marine structures, storage tanks, and pipelines. The cathodic protection was popular earlier for steel structures protection in a marine environment until it was used to protect reinforced concrete bridge deck recently.

A typical cathodic protection system used for reinforcement concrete structures consists of the following essential elements: the protected reinforcement, an anode, a power source, concrete surrounding the steel, a monitoring system, and cabling to carry the system power and monitoring signals, as illustrated in figure 6.4.

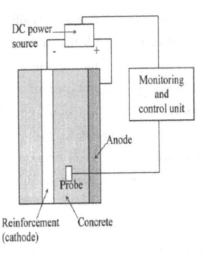

Figure 6.4 Typical cathodic protection systems

Installation of an anode into the reinforcement concrete system can go a long way in mitigating against further corrosion in the system. This can be achieved by removing the concrete down to the uppermost layer reinforcement to attach the anode. Finally, the exposed reinforcement and the anodes are embedded with sprayed concrete. The positive terminal of the power source is connected to the anode, and the negative terminal is connected to the reinforcement that acted as a cathode. The cathodic protection system contains two anodes. The first anode acts as a contact point and a power supply line for the second anode, which distributes the current over the surface of the structure when a certain amount of direct current is applied. The current flows through the electrolyte from the anode to the reinforcement, making the reinforcement become cathodic in relation to the anode. The pore water in the concrete contains alkalis and acts as an electrolyte, thus allowing the transfer of current from the anode to the reinforcement.

Epoxy-coated reinforcement was developed in the early 1970s in response to the need for better corrosion protection on reinforced concrete bridge decks. Epoxy-coated reinforcing steel is commonly used for reinforced concrete bridge decks around the world, and this procedure has saved billions of dollars in rehabilitation costs. Epoxy resins are thermosetting plastics that have good long-term durability in concrete and are resistant to solvents, chemicals, and rainwater. Epoxies have been widely used as both penetrating sealers and coatings in construction industry because they are resistive to chemicals and weather and also cure quickly.

The method of coating reinforcing steel with epoxy was adapted from the method used in the petroleum industry for marine steel structure coating. The bar is cleaned by blasting to remove mill scale and rust. The bar is then heated to temperature of around 230 degrees Celsius, which is essential for the application of epoxy powder, before it passes through an electrostatic spray that applies charged and dry epoxy powder to the steel. The epoxy melts flow and cures on the bars and then is quenched with a water spray bath.

Another cheaper way of protecting rebar is by coating it with zinc phosphate. Zinc phosphate gradually reacts with calcium cations and the hydroxyl anions in the cement pore water to form a stable hydroxyapatite layer. Corrosion inhibitors, such as calcium nitrite, may also add to the water mix before pouring concrete to prevent corrosion of the rebar. The nitrite anion is a mild oxidizer that oxidizes the soluble and mobile ferrous ions present at the surface of the corroding steel and causes it to precipitate as an insoluble ferric hydroxide. Solid stainless-steel reinforcement provides longer service lives than conventional reinforcement in a corrosive environment. However, the use of this method is limited as a result of high cost compared with other methods. A problem in utilizing stainless steel–clad reinforcement is the difficulty in bonding the cladding to the bars, and the exposed area of mild steel not covered with stainless material in concrete may be corroded.

Several techniques have been applied to improve the durability of reinforced concrete structures in the construction industry. A few of these techniques perform better than others. However, there is no single perfect method for all the environments. For the concrete structures built in an onshore environment, adequate concrete cover over usual steel reinforcement, and quality concrete with an appropriate water-cement ratio, is sufficient to guarantee the structures against corrosion damage. However, in the case of marine structures, additional protection and adequate concrete cover is generally required as a result of more aggressive conditions in a marine environment.

The determination of reinforcing steel's state of corrosion damage should be made before beginning with other construction measures. Special attention should be given to the concrete quality, depth of carbonization, chlorine ion profile, and moisture distribution.

CHAPTER 7

CATHODIC PROTECTION SYSTEMS

7.0 Introduction

The Proceedings of the Royal Society (1824) state that the science of cathodic protection began when Sir Humphrey Davy used iron anodes to protect the copper sheeting on the bottom of the British Navy's sailing ships. Sacrificial anodes made from iron attached to the copper sheath of the hull below the waterline dramatically reduced the corrosion rate of the copper.

Michael Faraday carried out extensive research on cathodic protection science and discovered the quantitative connection between corrosion weight loss and electric current. This finding laid the foundation for the future application of cathodic protection. Sir Humphry Davy experiment earlier carried out on impressed current cathodic protection on ships was unsuccessful due to lack of suitable current source and anode materials. However, cathodic protection application to steel gas pipelines began several decades ago and is more widely use till today.

7.1 Cathodic Protection

Cathodic protection (CP) is a technique used to control the corrosion of a metal surface by making metal the cathode of an electrochemical cell. The sacrificial metal then corrodes instead of the protected metal. Corrosion in aqueous solutions occurs by an electrochemical process. Anodic and cathodic electrochemical reactions occur simultaneously. There is no charge buildup on the metal as a result of corrosion because the rate of the anodic and cathodic reactions is always equal. Anodic reactions involve oxidation of metal to its ion, and for steel the following reaction occurs.

$$Fe \rightarrow Fe^{2+} + 2e \qquad \qquad \textbf{(7.1)}$$

The cathodic process entails reduction, and several reactions are possible. In acidic water, where hydrogen ions (H^+) are plentiful, the following reaction occurs.

$$2H^+ + 2e \rightarrow H_2 \qquad \qquad \textbf{(7.2)}$$

The rate of reactions in equations 7.1 and 7.2 can be changed by withdrawing electrons or supplying additional electrons to the piece of the metal, respectively. According to Le Chatelier's principle, if a dynamic equilibrium is disturbed by changing the conditions, the position of equilibrium moves to partially reverse the change. Thus, if we withdraw electrons from the piece of metal, the rate of reaction in equation 7.1 will increase to attempt to offset our action, and the dissolution of iron will increase, whereas reaction in Equation 7.2 will decrease.

Conversely, if we supply additional electrons from an external source to the piece of metal, the reaction in equation 7.1 will decrease to give reduced corrosion, and the reaction in equation 7.2 will increase. The second condition is applied to cathodic protection, and therefore to prevent corrosion, we need to keep on supplying electrons to the steel from an external source. It is known that the anodic and cathodic processes are indivisible, reducing the rate of the anodic process will cause the rate of the cathodic process to swell, as illustrated in figure 7.1.

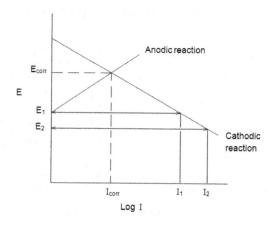

Figure 7.1 Kinetics of anodic and cathodic reaction

The previous description can be expressed in a quantitative manner by plotting the potential of the metal against the current densities of anodic and cathodic reaction rates. The corrosion current, I_{corr}, and the corrosion potential, E_{corr}, occur at the point of intersection where anodic and cathodic reactions rates are equal. When electrons are driven into metal, the process will make it more negative, and the rate of cathodic current increases to I_1, whereas the anodic dissolution of iron is significantly decreased at a potential E_1.

If the potential of the metal is further reduced to E_2 due to additional drive of current from an external source, the cathodic current will also increase to I_2. However, the metal will be considered to be overprotected if further protection of the metal is increased and more current is supplied from the external source. Excessive negative potentials can cause hydrogen evolution at the cathode surface on high-strength steels, leading to hydrogen embrittlement of the steel. This phenomenon will subsequently lead to the steel's loss of strength and also cause disbanding of insulating coating.

The basic principles on which cathodic and anodic protection operate are not completely the same. The two methods' operating philosophy in the course of protecting a given metal is presented in tabular form in table 7.1.

Table 7.1 Cathodic and anodic protection principles

S/N	Cathodic Protection	Anodic Protection
1	The principle is to force metal to be protected and to behave like a cathode	The principle is to increase the passivity of base metal by applying current in the direction that makes metal become more anodic
2	The cathodic protection involves reversing the flow of current between two dissimilar electrodes	The anodic protection involves suppression of anodic reaction by adjusting the potential of the more reactive metal
3	The cathodic protection can be achieved by the sacrificial method	The anodic protection can be achieved by the impressed current method

Cathodic protection (CP) systems are used extensively to prevent facilities from corroding, especially when the failure will have serious consequences such as loss of life and damage to property and the environment. CP systems are used to protect a wide range of metallic structures in various environments. Some of the common applications includes: pipelines and storage tanks, ship and boat hulls; offshore oil platforms and onshore oil well casings and metal reinforcement bars in concrete buildings and bridges. Appropriately designed and operated CP systems will significantly reduce the rate of corrosion, thereby extending the useful life expectancy of the protected facilities. However, for carbon and low-alloy steels, cathodic protection should be considered as a technique for corrosion control, rather than to provide coating alone against corrosion. Also, cathodic protection is not an alternative to corrosion-resistant alloys used for high-forbearance components such as subsea production systems.

Seawater pH and carbonate content are factors that affect the formation of calcareous layers associated with CP and the current needed to maintain the CP of bare metal surfaces. One of the disadvantages of cathodic protection is the formation of hydroxyl ions and hydrogen at the surface of the protected object. The presence of hydroxyl ions and hydrogen at the object surface often lead to disbanding of nonmetallic coatings due to chemical dissolution and electrochemical reduction processes at the metal-coating interface known as cathodic disbanding, as described in DNV-RP-B401.

Cathodic protection will cause formation of atomic hydrogen at the metal surface as the production of hydrogen increases exponentially toward the negative potential limit. The hydrogen atoms can either combine with metals to form hydrogen molecules or become absorbed in the metal matrix. Interaction of hydrogen atoms with the microstructure of components subject to high stresses initiates the growth of hydrogen-related cracks, known as hydrogen-induced stress cracking.

The main advantage of cathodic protection over other forms of anti-corrosion treatment is that it is applied simply by maintaining a direct current (DC) circuit, and its effectiveness may be monitored continuously. Specifying the use of cathodic protection may avoid the need to provide a "corrosion allowance" that may be costly to fabricate. Cathodic protection is used to prevent leakage where even a small leak cannot be tolerated for reasons of safety or environment because it is applicable to oil and gas pipelines. The benefit of cathodic protection for the protection of pipeline can hardly be overemphasized, as demonstrated in cathodic protection of a steel pipeline in Niger Delta illustrated in figure 7.2.

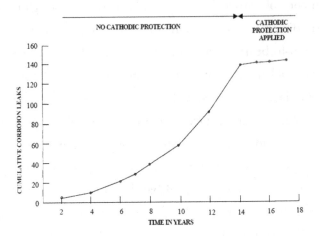

Figure 7.2 Pipeline corrosion leaks versus time

The graph shows the cumulative corrosion leaks versus time for the buried steel pipeline. It was observed that the corrosion leaks occurred at an ever-increasing rate prior to the application of cathodic protection in year fourteen, after which the corrosion rate becomes negligible. Cathodic protection may be achieved in one of two ways, by the use of an impressed current from an electrical source, or by the use of sacrificial anodes (galvanic action). The basic principle of cathodic protection is that metal dissolution is reduced through the application of a cathodic current, thereby reducing the metal corrosion potential by bringing the metal closer to an immune state. Cathodic protection is often applied to coated structures, with the coating providing the primary form of corrosion protection.

The CP current requirements tend to be excessive for uncoated systems. The use of coatings drastically reduces the CP current demand of the protection object and consequently the required anode weight. For controlled high-weight structures and long design life, the combination of a coating and CP is mostly cost-effective corrosion control. The use of coatings should be considered for the surfaces that are partly shielded from CP due to their geometrical effects. Also, for a large and complex structure,

MATTHEW OMOTOSO, PH.D

extensive use of coating is required to limit the overall current demand and to guarantee adequate current distribution. However, the application of coatings is not suitable for the parts of submerged structures.

Cathodic protection can be achieved in either of two ways based on the source of the current. The techniques are impressed current cathodic protection system and sacrificial anodes (galvanic action) methods. The impressed current is provided by an external current source and sacrificial anodes, with corrosion potential lower than the metal to be protected. To understand the action of sacrificial anodes for cathodic protection, it is quite necessary to have in mind the galvanic series of metals. When the tendency for metal to go into solution as metal ions increases, the metal becomes electronegative. Therefore, zinc, aluminum, and magnesium are used as sacrificial anodes to protect steel because they are more electronegative metals than steel.

7.2 Cathodic Protection System Selection

While selecting the most appropriate cathodic protection system for facilities' protection, the designer should first consider the size of the structure to be protected and the past project experience based on the operating and maintaining logistics for each type of cathodic protection system. Sacrificial cathodic protection systems are generally used for the protection of a well-coated facility, where protective current requirements are low and surface area of a protected facilities are relatively small. Thus, sacrificial cathodic protection systems are the best application for most of the subsea hardware due to their little maintenance cost. Impressed cathodic protection system is the most suitable type for the structures where protective current requirements and life requirements are high. The method is good for use over a wider range of soil and water resistivity and for protection of large uncoated areas, where relatively few anodes are required. However, both of the cathodic protection techniques (impressed cathodic and sacrificial anode) have their advantages and disadvantages, as spelled out later.

7.2.1 Advantages of Impressed Current

- Can be designed for a wider range of voltage and current applications
- Higher ampere-year can be obtained from each installation
- One installation can protect a more extensive area of the surface of the metallic structure
- Voltage and current can be varied to meet changing conditions such as increase protection of the surface coating
- Current required can be read and monitored easily at the rectifier
- Impressed currents system can be used for protecting bare or poorly coated surfaces of metallic structures

7.2.2 Disadvantage of Impressed Current

- Impressed current cannot be used without external source of power
- Installation is complex due to connection to external power source, rectifier and anodes
- The anode installation is complex and cannot be used for moving complex structures like ship where routing of cables from impressed current system could be problematic.
- Impressed current system requires frequent inspection and maintenance.

7.2.3 Advantages of Sacrificial Anode

- External power source is not required
- Installation is less complex because an external power source and rectifier are not required

- The system works very well when resistivity is low and the structure is well coated
- The anode system is easier to install on moving complex structures where routing of cables from impressed current system could present a problem

7.2.4 Disadvantages of Sacrificial Anode

- Current output is limited, and driving potential is also limited; hence the protection for bare steel area from each anode is limited
- Sacrificial anode system cannot be justified in water media when large surface areas of a poorly coated metallic structure require protection
- Installation can be expensive where a greater amount of anode material is required due to higher anode consumption rates
- Sacrificial anode system cannot respond to the additional bare area because its current and voltage are limited and cannot be varied
- Due to the buildup of algae, silt, and other debris on sacrificial anodes, the current output of the anode may be reduced
- Design basis must consider future changing conditions of structure surface, which is not considered in the design of sacrificial anodes, unlike the impressed current system
- Because there is no method to monitor the sacrificial anode system to determine its operation effectiveness except by taking structure-to-electrolyte potential measurements, many times this type of system is ignored, resulting in damage to the structure

7.3 Impressed Current Cathodic Protection System

Impressed current cathodic protection systems use the same elements as the galvanic protection system, but the structures are protected by applying current from the anodes. The anode and the structure are connected by an insulated wire. As for the galvanic system, current flows onto the structure from the anode through the electrolyte. The main difference between galvanic and impressed current systems is that the galvanic system relies on the difference in potential between the anode and the structure. On the other hand, the impressed current system uses external power source to drive the current. The external power source is usually a rectifier that changes input AC power to the proper DC power level, or DC supply from solar panels. Impressed current cathodic protection system anodes typically are high-silicone cast iron or graphite.

Managing electric current on pipelines is the best way to control corrosion because the current flows from the line to the surrounding soil, and electrolytes may lead to pipe corrosion damage. Even a small amount of current flowing away from line can be regarded as dangerous. One ampere discharging from a steel pipe surface for a year may lead to loss of nine kilograms of steel due to corrosion wastage. Therefore the objective of engineers is to make the pipe to be the cathode by making currents flow through the soil toward the pipe rather than away from it. To achieve this, external sources of current need to be set up to allow it to flow through the soil toward the pipe. This type of system arrangement is referred to as impressed current cathodic protection. The arrangement for protecting a buried pipeline is illustrated in figure 7.3.

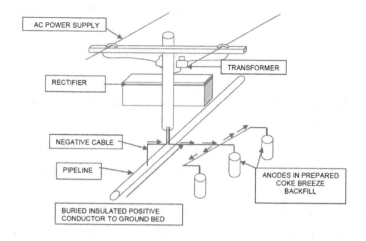

Figure 7.3 Impressed current cathodic protection station

Impressed current cathodic protection systems can be used to protect metal when coupled to the negative pole of a direct current source through a transformer-rectifier, while the positive pole is coupled to an auxiliary anode. Thus, the current needed for cathodic protection by impressed current is supplied from rectifier units. Because the driving voltage is provided by the current source, there is no need for the anode to be more active than the structure to be protected. All items to be protected shall be electrically connected and should have a welded or brazed connection to anodes.

In most cases, the public electricity supply grid is extensively used as a source of power supply to the CP transformer-rectifier. Solar cells, thermogenerators, and batteries for low-protection currents are used as a source of current in exceptional cases when there is no public electricity supply. In choosing the location for the anode installation and protection installation for local distribution and transmission lines, the following considerations are important and should be given in their order of consequence.

- Availability of a grid collection

- The electrical soil resistivity should be as low as possible in the area of the anode bed
- Sufficient distance from buried foreign installations to reduce their interference
- As little interruption to the site owner's property as possible
- Good access for vehicles
- For the stray current installations, a location that has as large a leakage area as possible for the stray current of the pipe network

The location of impressed current station should have guaranteed access at all time for inspection and maintenance. Any form of construction and damage to the location of the anode bed should be prevented. Where very high protection currents are involved, the length of cable should be made as short as possible between the protected object and the anodes to reduce major electric shock risk.

7.3.1 Transformer Rectifier

Transformer rectifiers are available for use in corrosion protection stations with various rates of direct current (DC) output. For storage tanks and short pipelines, it is about 10 watts, and for longer pipeline transformer rectifiers, output can be between 100 and 600 watts. A typical cathodic protection transformer rectifier produced by METEK, which is a solid-State, is shown in figure 7.4. It is also advisable that the rated current of transformer rectifiers be two times of the required current so that sufficient reserve is available to cope with the following future occurrences.

- Potential future enlargement of the installation
- A possible decrease in the coating resistance
- Increase in stray currents and other changes
- Sufficient reserve should be provided for the output voltages

Figure 7.4 Multiple output automatic rectifiers (AMETEK)

Transformer rectifiers must not be destroyed by high voltage and must be able to prevent an overshoot in the current supply grid. There should be a limiting action on the pipe and anode voltage. However, a cascade of chokes, overvoltage arresters, and a condenser between bridge rectifier and output prevent these incidents from happening. Control of current output of rectifier can be manual or automatic. In the case of manual control, either current or voltage will be used as control parameters. While using automatic control, it shall be based on potential reading from fixed reference electrodes. The alarm will be tripped off when there is more or less voltage/current or high protection potential on anodes. The magnitude of potential and current required to protect a given structure will depend on the surface area of the structure. The polarized potential charge at the drain point should be approximately 1.2 volts; this limits the surface area that can be protected from a single drain point, and it also determines the distance between rectifiers in a several drain-points system. Transformer rectifiers should have an ammeter to indicate the current and a high-resistance voltmeter to signify the potential at the protection station. Failures can occur in rectifier-groundbed systems, and such failures include the following.

- Permanent or temporary power supply failure
- Failure in rectifier stacks, circuit breaker, or auxiliaries
- Cable disconnection to the anodes or protected structure
- Total or partial consumption of anodes
- Deterioration of pipe coating, resulting in an increase in current demand of the system

The cathodic protection stations are often operated under continually changing conditions. These conditions may include stray current interference, changes in grounding resistance, and fluctuation in the required current due to variation in the aqueous level and oxygen access as applicable to the storage tank. Therefore, in case any of the conditions occur, it is recommended that the transformer rectifier be provided with an electrical control circuit to keep the system potential and current constant. To achieve this objective, equipment such as potentiostats and galvanostats can be provided for the system potential and current controlling, respectively.

The rectifier is selected based on the outcome of voltage output (V_{rec}) and the power, which can be calculated by equations 7.3 and 7.4, respectively.

$$V_{rec} = (I) \times (R_T) \times (150\%) \tag{7.3}$$

$$P = (V_{rec}) \times (I) \tag{7.4}$$

In these equations, I is the protective current in ampere, (R_T) is the circuit resistance, 150% is a factor to allow for aging of the rectifier stacks, and P is the required power. Several rectifiers are available commercially. However, only the rectifier that satisfies the minimum requirements of voltage output is recommended for use. In addition to the general rectifiers being marketed, a solar cathodic protection power supply for direct current power may be considered for the remote sites with no electrical power.

While recommending a solar cathodic protection power supply, the following factors should put to consideration.

- The cost of the solar cathodic protection power in unit price per watt of continuous power
- The initial cost of solar cathodic protection power supply is higher than selenium rectifier operated by AC power
- Continued maintenance cost of keeping the solar panel free of dirty deposits

The groundbed header cable resistance, which is available in ohms for copper of 35 mm², is 0.63 (ohm/km) and can be calculated using equation 7.5, where L_u is the useful length of cable and L_c is equal to the cable resistance.

$$R_W = (Lu) \times (.Lc) \tag{7.5}$$

The number of anodes (N) needed to meet maximum anode groundbed resistance requirements is calculated using equation 7.6, where Ip is total protection current, Ai is the anode surface area, and Ic is equal to anode current limit. For a silicon anode, that is 10.76 A/m².

$$N = \frac{Ip}{(Ai) \times (Ic)} \tag{7.6}$$

The anode resistance is determined using equation 7.7, where R_a is the anode resistance, ρ is the soil resistance in ohm-centimeters, K is the anode shape factor (table 7.2), N is the number of anodes, L is the length and diameter of the anode backfill column in feet, P is the paralleling factor (table 7.3), and S is the center-to-center spacing between anode backfill columns in feet.

$$R_a = \frac{(\rho)\times(K)}{(N)\times(L)} + \frac{(\rho)\times(P)}{S} \qquad (7.7)$$

Table 7.2 Impressed current anode shape functions

L/d	Anode Shape Function (K)	L/d	Anode Shape Function (K)
5	0.0140	20	0.0213
6	0.0150	25	0.0224
7	0.0158	30	0.0234
8	0.0165	35	0.0242
9	0.0171	40	0.0249
10	0.0177	45	0.0255
12	0.0186	50	0.0261
14	0.0194	55	0.0266
16	0.0201	60	0.0270
28	0.0207		

Calculation of structure-to-electrolyte resistance can be estimated using equation 7.8, where R_C is the coating resistance in ohms per square meters and A_p is the coated pipe area in square meters.

$$R_C = \frac{R}{A_p} \qquad (7.8)$$

Total circuit resistance, R_T, is obtained using equation 7.9

$$R_T = R_a + R_W + R_C \qquad (7.9)$$

Table 7.3 Impressed current anode paralleling factors

Anode Quantity (N)	Paralleling Factors (P)	Anode Quantity (N)	Paralleling Factors (P)
2	0.00261	14	0.00168
3	0.00289	16	0.00155
4	0.00283	18	0.00145
5	0.00268	20	0.00135
6	0.00252	22	0.00128
7	0.00237	24	0.00121
8	0.00224	26	0.00144
9	0.00212	28	0.00109
10	0.00201	30	0.00104
12	0.00182		

7.3.2 Anode and Backfill Materials

The shape of anode for impressed current CP is very important for efficiency, but in most cases the cylindrical shapes are considered to be the best. The anode material will determine the ratio of the length and the diameter. The rapid consumption of the anode end due to extra current discharge in the cylindrical end is known as end effect. This development creates a problem for the easy connection of conductor wire to the anode end.

Selection of anode material needs to be given due consideration because anode is expected to have a low consumption rate irrespective of environment, low polarization levels, high electrical conductivity, and low resistance at the anode-electrolyte interface. The lowest possible grounding resistance should be considered for impressed current anode in order to keep down the electric power and consequently the operating costs. The desirable properties of an "ideal" impressed current anode material are as follows.

- Low electrical equivalents
- Uniform composition
- Shape that assures complete consumption
- Adequate mass that is enough for facilities life span
- Surface area that reduces circuit resistance
- High mechanical integrity to minimize mechanical damage during installation, maintenance, and service use
- High resistance to abrasion and erosion
- Ease of fabrication into different forms
- Low cost relative to the overall corrosion protection scheme

In practice, among the anodes that satisfy the above conditions are anodes of compressed and baked graphite and high-silicon cast iron. Anodes are characterized by material and geometry. The common sacrificial anode materials are either aluminum or zinc based. Aluminum-based anode materials are favored due to their higher electrochemical efficiency and light weight compare with zinc-based materials. The geometry of anodes can be cylindrical shape, slender standoff, or bracelet.

The type of anode to be used to protect a structure is normally determined by the designer at the beginning of a project after due consideration to the following factors.

- Anode utilization factors
- Installation restriction
- Anode weight
- Current drag force
- Manufacturing and installation cost
- Obstruction to other operations

Improvement to the performance of impressed current anodes installed underground should be surrounded by a pack of low-resistance carbon such as coke breezed or crushed coke, as illustrated in figure 7.5.

MATTHEW OMOTOSO, PH.D

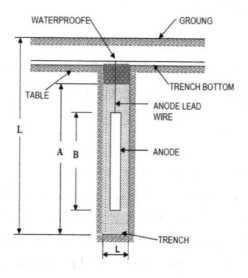

Figure 7.5 Anode installed underground

The aim of the backfill materials is to provide good electrical contact between the anode and the surrounding earth. Backfill material assists to lengthen the life span of an anode by allowing the current leaving the anode to be electrically conducted to the earth. Also, backfill material increases the effective surface area of the positive circuit that is in contact with the earth to save the electrical cost by lowering the resistance of the anode bed, eliminating gas blockage, and drying tendencies. Coke breeze is a carbonaceous backfill widely used to enhance the performances of impressed current groundbed. Desirable properties of a good backfill material include the following.

- Low cost
- Low resistivity
- High density
- Low consumption rate
- Good natural compaction
- Easily fluidizable in water

- Good settling characteristics
- Good gas permeation.

Coke breeze is a carbonaceous backfill widely used to enhance the performances of impressed current groundbed

7.3.3 Anode Cable

For an impressed current system, it is important that the anode cable, splices, and collections in the anode system are completely insulated from surrounding medium to avoid corrosion and damages. Therefore, anodes are recommended to be placed near the rectifier to reduce the anode cable length. Also, anode and drain cables should be carefully selected in order to transmit the maximum rectifier current to the structure.

Economics are important in choosing a cable and may indeed be a controlling factor to determine the total annual cable cost. Kelvin's economic law can be used, as shown by equation 7.10, where T is total annual cost in dollar per year, I is total protection current in amperes, R is cable resistance in ohms per 1,000 feet (305 meters), L is cable length in feet, P is cost of electrical energy in Farads per kilowatt-hour, E is the rectifier efficiency expressed as a percentage, and S is the cable's initial cost in dollars per foot.

$$T = \frac{(0.0876) \times (I^2) \times (R) \times (L) \times (P)}{E} = (0.15) \times (S) \times (L) \qquad \textbf{(7.10)}$$

Based on manufacturer specification of anode cables' resistance, the groundbed resistance for header cable is calculated using equation 7.11, where L is cable length in feet.

$$R_W = \left(\frac{ohms}{unitLength} \right) \times L \qquad \textbf{(7.11)}$$

7.4 Sacrificial Cathodic Protection System

In the sacrificial cathodic protection application, the naturally occurring electrochemical potentials of different metals are used to provide protection. Sacrificial anodes are coupled to the structure, which is under protection, and conventional current flows from the anode to the structure once the anode is more "active" than the structure. As the current flows, all the corrosion occurs on the anode in order to offer protection from corrosion to the structure. Because the most active metal (anode) supplies current, it will gradually dissolve into ions in the electrolyte, and at the same time it will produce electrons, which the least active (cathode) will receive through the metallic connection with the anode. The result is that the cathode will be negatively polarized and therefore be protected against corrosion.

On the active side of the galvanic series for common metal, the metals such as zinc, aluminum, and magnesium appear. These metals and their alloys are the most commonly used as sacrificial anodes. Sacrificial cathodic protection systems employ sacrificial anodes attached to the steel. The anodes are connected to the steel by bars or cables. The state-of-the-art sacrificial CP systems involves the use of galvanic aluminum anodes, as in figure 7.6.

Anode

**Figure 7.6 Anodes attached to subsea
Christmas tree (FMC Technologies)**

Advice on dimension and distribution of anodes is given in the relevant codes and regulations. Sacrificial cathodic protection system is preferred on moving complex structures, where routing of cables from impressed current system could present a problem and the installation is less complex because an external power source and rectifier are not required. The detriment of this system is that current output is limited and driving potential is also limited; hence, the protection for bare steel area from each anode is limited.

Due consideration should be given to the effects of dissimilar metal, such as stainless-steel parts alongside carbon-steel parts in the subsea hardware. Carbon steel will act as sacrificial anodes in the presence of the noble metal as a result of the galvanic potential established between the metals. Therefore, as a general rule, avoid the use of the dissimilar metals on exterior surface exposed to seawater. In case the use of the dissimilar metals cannot be avoided, the following mitigation measures are recommended.

- Provide an impressed current source to reverse the detrimental galvanic potential between dissimilar metals.
- Apply paint over exposed noble metal surfaces; this device is to have a little cathode and a larger anode to slow down the galvanic potential between dissimilar metals.
- Provide electrically insulated connection points, such as bolted flange connections, with coatings or dielectric shields to break the galvanic circuit.
- Increase the design current density (i_c) to offset the effect of higher galvanic potential between the dissimilar metals.

7.4.1 Cathodic Protection Environmental Parameters

The cathodic protection system is to be designed with regard to environmental conditions. The most important seawater factors affecting a cathodic protection system design, which is also interrelated and varies with geographical location, are dissolved oxygen content, sea currents, temperature, marine growth, and seawater salinity. These parameters are considered for the establishment of cathodic current density demand to attain and maintain cathodic protection of a system for a given period of time.

Where sacrificial anodes are used, the exact location and distribution of anodes should be identified on the protected structures. However, the design should confirm that the installation of anodes is not impeding subsea intervention operations. Anodes are to be installed in the water-submerged zone, where possible, and are not advised to be installed below the mud line for effective performance. For the pipeline that is protected by a sacrificial cathodic system, anode spacing is recommended to not exceed two hundred meters. The quantity of anodes should be increased by a factor of about two for the first five hundred meters from any installations to compensate for possible current leakage to other surrounding facilities.

7.4.2 Current Density Requirements

Current density is defined as cathodic protection current per unit surface area that requires maintaining the protection potential that is dependent on local conditions. It is not practicable to establish the accurate relationship between those environmental factors and cathodic current demand. Consequently, the design current densities have been defined based on depth and climatic zones, as shown in tables 7.4 and 7.5, respectively.

The increased availability of oxygen, water flow or turbulence, and the pH of environment will directly increase the current density required to protect a marine structure. The presence of coatings, marine fouling, and calcareous deposits will have a profound effect on current density. For bare steel surfaces buried in sediments, the design current density for all three stages, irrespective of geographical location and depth, is to adopt 0.020 A/m^2, per RP B401 (2005).

Table 7.4 Current densities in seawater for bare metal, from RP B401

Depth (m)	Design Current Densities (Initial/Final) in A/m^2							
	Tropical > 20°C		Subtropical 12–20° C		Temperate 7–11° C		Arctic < 7° C	
Stages	Initial	Final	Initial	Final	Initial	Final	Initial	Final
0–30	0.150	0.100	0.170	0.110	0.200	0.130	0.250	0.170
>30–100	0.120	0.080	0.140	0.090	0.170	0.110	0.200	0.130
>100–300	0.140	0.090	0.160	0.110	0.190	0.142	0.22	0.17
> 300	0.180	0.130	0.200	0.150	0.220	0.170	0.220	0.170

Table 7.5 Mean current densities in seawater for bare metal, from RP B401

Depth (m)	Design Current Densities in A/m²			
	Tropical > 20° C	Subtropical 12–20° C	Temperate 7–12° C	Arctic < 7° C
0–30	0.070	0.080	0.100	0.120
>30–100	0.060	0.070	0.080	0.100
>100–300	0.070	0.080	0.090	0.110
> 300	0.090	0.100	0.110	0.110

7.4.3 Initial Design Current Density

Initial design current density is defined as current density required to cause polarization of an initially exposed bare metal surface. The current density is typically higher than the final current density due to protective calcareous scale formation during the final stage, which reduces the final current quantity. The values for initial current density based on the climatic zone and water depth are presented in table 7.4.

7.4.4 Average Design Current Density

This is the estimated cathodic current density needed when the cathodic protection system has attained steady state protection potential. The average design current is to compute the minimum mass of anode material desired to maintain cathodic protection system during the design life, and this quantity is conservative. The values for average current density per each climatic zone and water depth are presented in table 7.5.

7.4.5 Final Design Current Density

The current density required to maintain the protection potential depends on local conditions. Increased availability of oxygen at the surface of metal will directly increase required current density. The value of this current takes into consideration the additional current density required to repolarize the metal surface due to partial and occasional damages. The values of final current density for each climatic zone and water depth are presented in Table 7.4. With regard to the three stages of design current density, and for the surface with operating temperatures exceeding 250°C, the current density shall be increased by 0.001 A/m^2 per degree prior to the effect of coating factors (RP B401 2005).

7.4.6 Coating Quality and Breakdown Factor

The provision of insulating coating for the corrosion-protected structure greatly reduces the current demand for cathodic protection. The conjoint use of coatings and cathodic protection takes advantage of the most attractive features of each method of corrosion control. Coating provides the bulk of protection, and cathodic protection provides protection to flows in the coating. As the coating degrades with time, the activity of the cathodic protection system develops to supplement the deficiencies in the coating. Therefore, the combination of coating and cathodic protection generally provides the most economic corrosion protection system.

However, when higher potential is applied to a coated structure, it has a damaging effect on the coating. The hydrogen evolution on the structure surface destroys the bond between the coating and the structure. Therefore, with increases to the coating conductance, the rate of coating deterioration increases rapidly as polarized potential is increased.

Table 7.6 Coating breakdown factors, from RP B401

Depth (m)	Recommended *a* and *b* Value for Coating Categories I, II, III		
	I (a = 0.10)	II (a = 0.05)	III (a = 0.02)
0–30	b = 0.10	b = 0.025	b = 0.012
> 30	b = 0.05	b = 0.015	b = 0.008

The coating breakdown factor (f) can be defined as an estimated reduction in cathodic current density due to the application of an electrically insulated coating, where $f_c = 0$; this shows that coating is 100 percent electrically insulating. Therefore, the cathodic current density is decreasing to zero, where $f_c = 1$; this indicates that the coating has no protective properties, and the cathodic current density of an initial coated surface should be the same effect as a bare steel surface. Also, the coating breakdown factor is a function of coating properties and operational parameters. The coating properties (f_c) refer to coating material, surface preparation, and method of application, which is expressed mathematically with equation 7.12.

$$f_c = a + b \times t \tag{7.12}$$

Here, t is coating lifetime in years, and a and b are constant depending on coating properties and depth, as shown in table 7.6. The coating breakdown factor f_{cm} and f_{cf} are to be calculated using equations 7.13 and 7.14, respectively, as stipulated in RP B40 (2005).

$$f_{cm} = a + \frac{b \times t_f}{2} \tag{7.13}$$

$$f_{cf} = a + b \times t_f \tag{7.14}$$

If the design life (t_f) of the cathodic protection system is longer than the actual calculated life of coating system, f_{cm} may be determined by equation 7.15. Where the estimated value according to 7.13 and 7.15 exceeds 1, f_c equal to 1 shall be adopted.

$$f_{cm} = 1 + \frac{(1-a)^2}{2.b \times t_f} \tag{7.15}$$

However, the category of coating and estimating area percentage of bare metal exposure, which may occur from coating damage, needs to be recognized. The coating's resistance decreases greatly with age and directly affects structure-to-electrolyte resistance for design calculations. But the coating producers typically supply coating resistance values among the chemical data provided by the coating manufacturer.

7.4.7 Current Demand

Current demand is the quantity of current required to protect a structure against corrosion. The value is directly proportional to structure surface area, coating breaking down factor, and design current density. The value represents mathematically with equation 7.16.

$$I_c = (A_c) \times (f_c) \times (i_c) \tag{7.16}$$

Here, I_c is the current density (initial, average, final), A_c is the individual area, f_c represents coating breakdown factor, and i_c is the design current density.

The decisive part of design calculations for cathodic protection systems is the amount of current required per square meter to change the structure's potential to (-0.85) volts, known as current density. The current density required depends on the structure's surface condition. A well-coated structure will require a very lower current density compared to an uncoated surface.

The current density required per square meter can be determined by an actual test on the existing structures using a temporary cathodic protection setup, theoretical calculation based on coating efficiency, and estimation of current requirements using tables based on field experience.

7.4.8 Anode Estimation

The total anode mass (M_t) essential for maintaining cathodic protection throughout the design life of a structure can be estimated using equation 7.17.

$$M_t = (D) \times (I_R) \times \left(\frac{m}{u} \right) \tag{7.17}$$

Here, D represent design life, I_R required intensity (average), u utilization factor, m consumption rate, given by $m = \frac{8760}{\varepsilon}$, ε electrochemical efficiency. The cathodic protection system is always designed to protect a structure for a given number of years. To meet this lifetime requirement, the number of anodes (Q) is estimated using equation 7.18.

$$Q = \frac{M_t}{Anode\ Unit\ \text{weight}} \tag{7.18}$$

Determination of anode quantity by current density is essential for the structure protection for a given life span. The design of a complete cathodic protection system is calculated using this equation.

$$Q = \frac{I_{required}}{\text{Anode current output}} \tag{7.19}$$

The calculation for anode current output at a given condition for a cathodic protection system can be expressed as follows.

$$I = \frac{\Delta E}{\text{Anode resistance}}, \quad \Delta E = E_c - E_a \tag{7.20}$$

E_c is the protective potential, and E_a is the alloy potential

The anode utilization factor is defined as the fraction of anode material that can be utilized for a design purpose. When the anode is enthused further than the utilization factor, the anode performance can no longer be guaranteed as a result of the increase in the anode resistance and the loss of support material. In the relevant design codes, the factor is recommended to be between 0.8 and 0.9 depending on anode shape and sizes.

7.4.9 Anodes Resistance

The value for short, flush-mounted bracelet and flush-mounted shapes anode resistance at initial and final conditions are calculated using equation 7.20 and 7.21, respectively.

$$R_0 = \frac{0.315\rho}{\sqrt{A}} \quad (7.20) \quad R_F = \frac{0.315 \times 30}{\sqrt{l \times b}} \tag{7.21}$$

Here, ρ is the environmental resistivity, A is the anodes exposed surface area, l- is the anodes length, and b- is the anodes length and width. The anode resistance for a long anode at initial and final conditions can be estimated based on equations 7.22 and 7.23, respectively.

$$R_0 = \frac{R}{L + W} \tag{7.22}$$

$$R_F = \frac{R}{L_f + W_f} \tag{7.23}$$

Here, L anodes length at initial condition, W anodes width at initial condition, L_f anodes length at final condition, and W_f anodes width at final condition.

7.5 Cathodic Protection Design Considerations

There are primary considerations that the design engineer should bear in mind while carrying out the design of a cathodic protection system, whether sacrificial anode or impressed current systems. These considerations include efficient distribution of sufficient current to achieve cathodic protection. Due to limited range of voltages available, the desired current becomes selection of the electrical circuit resistance. The cost savings of cathodic protection design, as well as the installation procedures, have attracted cathodic protection designers for making a choice out of the two systems.

A sacrificial anode system does not require an outside power source because the protective current is generated by the electrochemical reaction between the metals. This form of cathodic protection system is generally recommended for a well-coated structure with minimum chance of the anode being damaged during the installation's useful life. The relatively low and normally well-distributed current output results in a more constant current density on the protected structure, which minimizes overprotected and wasted current. The sacrificial anode is efficient, and the low current output reduces the possibility of interference.

The impressed current cathodic protection systems can offer significant economic advantages over galvanic anodes on certain types of facilities that require a cathodic protection retrofit. However, these advantages are not always there due to poor system performance with the negative economic repercussions. The impressed current system makes use of anodes that are made of durable materials and resist electrochemical wear. Rectifier supplies current to the system. The system often requires routine maintenance because it consists of power supply and a larger number of electrical connections. However, unlike sacrificial cathodic protection, an impressed current system effectively protects large and bare structures because of the system's greater current capacity.

Impressed current cathodic protection makes economic sense where there is large current requirement and significant water depth. These factors generally lend to using impressed current. These factors are interdependent to a large degree, though not always. For instance, impressed current may be attractive for a large structure in shallow water due to high current requirements and difficulties associated with installation of sacrificial anodes.

On the other side, a very deep structure inside the sea, such as a suction pile, may have a very low subsea current requirement due to extensive use of subsea coatings, which means impressed current is probably not a good option. As a general rule of thumb, impressed current systems begin to look quite attractive when current requirements exceed 400–500 amperes and when water depth exceeds 150 meters. Some structures may satisfy the mentioned criteria and yet may not be candidates for an impressed current if there is no available space to install the transformer rectifier and if there is risk of interference with other cathodic protection systems.

The extremely intricate structures make installation of impressed current not feasible. Therefore, moving vessels or structures that use very high strength steels may prevent the use of impressed current for corrosion protection. Also, there may be safety concerns for the system as a result of sport divers, fishermen, and endangered wildlife activities. The availability of trained personnel to operate and maintain the system is vital prior to the recommendation of impressed current due to the intricacy of the system. Therefore, before opting for sacrificial cathodic protection system or impressed current system, certain preliminary information should be reviewed with regard to the structure characteristics and the environments, as listed in table 7.7.

Table 7.7 Choice of cathodic protection systems

S/N	Activities	Impressed Current	Sacrificial Cathodic
1	Capacity to design for a wider range of voltage and current	✓	X
2	Requirement of external electrical power	✓	X
3	Simple installation procedure (comparing the two systems)	X	✓
4	Capacity for bare metal protection	✓	X
5	Opportunity for current monitoring	✓	X
6	Low maintenance cost (comparing the two systems)	X	✓
7	Capacity for wider metal area protection	✓	X
8	Installation of a limited number of anodes	✓	X
9	Installation of a transformer rectifier	✓	X
10	Requirement of right-of-way	✓	X

7.6 Cathodic Protection Design Codes and Standards

In the course of cathodic protection system design, adequate references should be given to specialized documents (specification and manuals) that are based on specific expertise. Other documents, such as conference proceedings, technical journals covering particular design aspect, and textbooks on corrosion control with special references, are highly recommended. A typical design basis document for cathodic protection design usually specifies criteria applicable for the structure protection with special reference to the design metallurgy, environmental factors, and operating requirements, so that the appropriate cathodic protection system can be adopted to be either sacrificial anode or impressed current protection system.

There are several codes and standards available for how cathodic protection systems should be designed, built, and operated. The design shall follow the standard set's recommended practices. There are several cathodic protection design codes and standards worldwide, and some of them are listed here.

- Det Norske Veritas Industri Norge As (Recommended Practice RP B401, Cathodic Protection Design)
- API Recommended Practice 1632, "Cathodic Protection of Underground Petroleum Storage Tanks and Piping Systems"
- NACE RP 0169, "Standard Recommended Practice: Control of External Corrosion on Underground or Submerged Metallic Piping Systems"
- NACE RP 0285, "Standard Recommended Practice: Corrosion Control of Underground Storage Tank Systems by Cathodic Protection"
- NACE Test Method TM 0497, "Measurement Techniques Related to Criteria for Cathodic Protection on Underground or Submerged Metallic Piping Systems"
- STI R892, "Recommended Practice for Corrosion Protection of Underground Piping Networks Associated with Liquid Storage and Dispensing Systems"
- STI-R-972, "Recommended Practice for the Installation of Supplemental Anodes for STI-P3 USTs"
- UL 1746, Standard for Safety: "External Corrosion Protection Systems for Steel Underground Storage Tanks"

An important element in designing a cathodic protection system is the structure's physical dimensions (e.g., length, width, height, and diameter). These data are used to calculate the surface area of the structure to be protected. The installation drawings must include sizes, shapes, material type, and locations of parts of the structure to be protected. Due consideration should be given to the structure sections in the corrosion locality zones for the appropriate selection of design current density values. Current density and area of structures are imperative for the required protecting current calculation and subsequent estimation of anode quantity for a given design life.

The activities during the installation phase include anode installation and probably isolation of some existing structures due to some reasons based on the decision taken during the design phase. A cathodically protected structure must be electrically connected to the anode. In some cases, parts of the structures are electrically isolated by insulators to eliminate possible interference with the existing cathodic protection systems. To comply with environmental regulations and industry standards, close-interval corrosion survey and preventive maintenance are required for all installed cathodic protection systems. The purpose of this preservation is to ensure that adequate cathodic protection is maintained over the entire protected structure. The sequence for the carrying out of either sacrificial cathodic protection or impressed current is represented in a flowchart in figure 7.7.

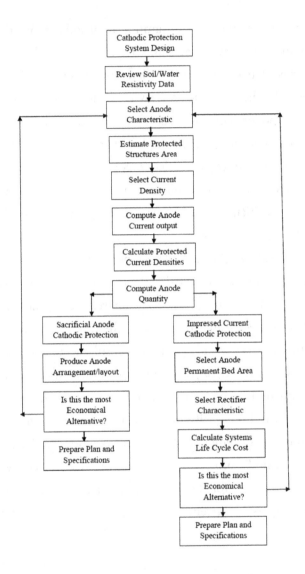

Figure 7.7 Cathodic protection system design sequence

7.7 Limitation of Cathodic Protection

Application of cathodic protection should be considered as a method for corrosion control, rather than providing full protection for the structure. Therefore, the technique should not be allowed to replace the selection of highly corrosion-resistant materials during the project design stages. Also, cathodic protection system may not be very effective under conditions that support harsh bacterial corrosion because it is applicable to the shielded portion of structures such as disbanded coating, coating holidays, and crevices. Negative effects of cathodic protection are formation of hydrogen ions and hydrogen at the surface of the cathode or protected structure, which may cause disbanding of coatings. Cathodic protections also cause the liberation of hydrogen atoms at the metal surface, and part of this hydrogen becomes ingressed into the metal medium. In addition to the tensile load on the structure, this hydrogen may cause hydrogen-induced stress cracking as a result of the decrease in the metal ductility and load-bearing capacity.

Special attention is also essential to prevent corrosion damage before structure installation, and the period required for cathodic protection system needs to energize. Also, for a vital structure that needs to withstand high pressure and a long-design life span, extra thickness may be provided as corrosion allowance. Corrosion allowance is an extra thickness of metal beyond that which is needed for strength. The parameter is normally established by the end user and is somewhat based on personal preferences and industry tradition. Taking into consideration the techno-economics and reliability of the structure, corrosion allowance is the better option. Thus, application of corrosion allowance is always considered as a reliable corrosion protection in the splash zone. The rate of corrosion increases proportionally with the age of the structure, and it is highly recommended to combine corrosion protection techniques, such as corrosion allowance, with surface treatment and cathodic protection systems.

CHAPTER 8

CATHODIC PROTECTION DESIGN EXAMPLES

8.0 Introduction

8.1 Suction Pile Cathodic Protection

Example -1

Twelve suction piles are proposed to be installed in a water depth of 1,200 meters with mooring chains to anchors Floating Production, Storage and Offloading (FPSO) system in offshore Nigeria. The suction pile structure is equipped with several internal stiffeners and padeye for lifting. The pile diameter is 6.50 meters and 21 meters long. Carry out cathodic protection design for the suction piles using aluminum as sacrificial anodes. Only the outside surfaces of the pile in the submerged and buried zones will be protected. The design life of the cathodic protection system is considered to be twenty-five years. The calculation method should be in accordance with the "DNV Recommended Practice RP B 401 Cathodic Protection Design 2005."

Codes and References

- Recommended Practice RP B 401 Cathodic Protection Design 2005 Det Norske Veritas Industri Norge AS
- FPSO PILE General structure—Drawing list

Anode Characteristics

- One type of sacrificial anode alloy will be used
- Aluminium anodes for protecting the outside of the pile (submerged and buried)
- Aluminium anodes will be the short, flush-mounted type

Table 8.1 Showing anode characteristics

Alloy type	Anode Length (mm)	Anode width (mm)	Anode weight (mm)	Net weight (kg)
Aluminium	560	220	180	65

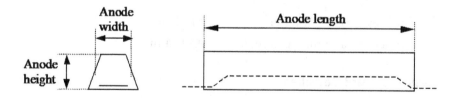

Figure 8.1 Anode diagram

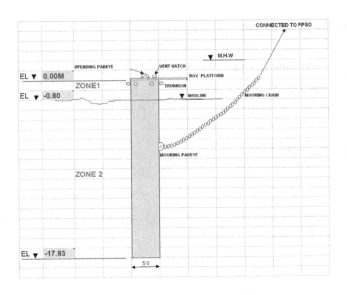

Figure 8.2 Suction pile with zone classification

Surface Areas Calculations

The following surface areas have been calculated for the suction pile, padeye, and mooring chain for the cathodic protection. For the detailed area calculations, see tables 8.4, 8.5, and 8.6.

Surfaces area in sea water:	81.41 m²
Surfaces area in mud external:	430.74 m²

Design Current Densities

For the depth 0–30 meters, the following current densities have been chosen for the submerged zone taken from table 7.4 and 7.5 with due consideration to the the project climatic region.

Initial:	200 mA/m²
Mean:	100 mA/m²
Final:	130 mA/m²

The mud zone (external surfaces) is 20 mA/m² according to paragraph 6.3.8 of RP B401 (2005).

Anodes Characteristics

Aluminium base anodes and anode dimensions were chosen from an economics point of view and design experience.

-    Cross section: 22×18 cm²
-    Total length: 56.0 cm
-    Density: 2.75 kg/dm³
-    Seawater resistivity: 30.cm

Anode Initial Exposed Surface Area

Anode initial surface area is calculated based on anode characteristics.

$$A_0 = 2(22 \times 18 + 56 \times 18) + 22 \times 56 = 4040 \text{ cm}^2$$

Anode Final Exposed Surface Area

It is assumed that the final exposed surface area is equivalent to the initial surface area of the anode facing the surface to be protected.

$$A_F = 22 \times 56 = 1232 \text{ cm}^2$$

Anode Resistance

Resistance at initial stage: $R_0 = \dfrac{0.315\rho}{\sqrt{A}}$ **(8.1)**

with ρ = environmental resistivity, A = the exposed anode surface area.

$$R_0 = \frac{0.315 \times 30}{\sqrt{2(22 \times 18 + 56 \times 18) + 22 \times 56}} = 0.149\Omega$$

Resistance at final stage: $R_F = \frac{0.315\rho}{\sqrt{l \times b}}$ (8.2)

with l = the anodes length, b = the anodes length and width.

$$R_F = \frac{0.315 \times 30}{\sqrt{56 \times 22}} = 0.269\Omega$$

Anode Current Output

$$I = \frac{\Delta E}{R}$$ (8.3)

$$\Delta E = E_c - E_a$$ (8.4)

E_c = protective potential = -0.800 V

E_a = alloy potential = -1.05 V

At initial condition: $\frac{0.25}{0.149} = 1.678$ A

At final condition: $\frac{0.25}{0.269} = 0.929$ A

Coated Surfaces

The section of the pile above the mudline is painted. The paint coating has been considered as category II according to DNV classification.

Coating Breakdown Factor

$$f_{cm} = a + \frac{b \times t_f}{2}$$ (8.5)

$$f_{cf} = a + b \times t_f \tag{8.6}$$

Initial $f_c = a$

t_f = design life

a and b for category II coating, as stipulated in table 8.2

$a = 0.05$ and $b = 0.025$

Table 8.2 Coating breakdown factor summary

Category II	Initial $f_c = 0.05$	Average f_c = 0.36	Final $f_c = 0.68$

Protection Current Densities

The quantity of current required to protect a structure against corrosion is provided in table 8.3.

$$I_c = (A_c) \times (f_c) \times (i_c) \tag{8.7}$$

I_c is the current density (initial, average, final), A_c is the individual area, f_c represents coating breakdown factor, and i_c is the design current density.

Determination of Anode Quantity

The total anode mass M_t required for maintaining cathodic protection throughout the design life is given by equation 8.8.

$$M_t = D \times I_R \times \frac{m}{u} \tag{8.8}$$

D = design life (25 years)

I_R = required intensity, average (tables 8.4, 8.5, and 8.6)

u = utilisation factor; a value of 0.80 has been taken for all the anodes

$$m = \text{consumption rate} = \frac{8760}{\varepsilon} = 4.38 \qquad (8.9)$$

ε = electrochemical efficiency; a value of 2,000 Ah/kg for aluminium-base anodes has been taken

Required Current Intensity

The required current intensity summary presented in table 8.3 is taken from tables 8.4, 8.5, and 8.6.

Table 8.3 Required Current Intensity Summaries

Protected Zone	Average	Initial Stage	Final Stage
External surfaces (submerged & buried)	11.55A	9.43A	15.81A

Total Mass of Anodes

$$M_t = (D) \times (I) \times (R) \times \left(\frac{m}{u}\right) \qquad (8.10)$$

$$M_t = 25 \times 11.55 \times \frac{4.38}{0.80} = 1581 kg$$

Table 8.4 Cathodic Protection Calculation Table at Initial Stage Current Required for FPSO Mooring Pile

Component Description	Net Quantity	Unit	Surface to Protection (Per Unit Length or Surface)	Unit	Surface Protection	Unit	Current Density mA/m²	Coating Breakingdown Factor	Current Required A
OUTSIDE SURFACE									
1. Surface In Sea water									
Head Plate	33.91	m²	1.00	m²/m²	33.91	m²	200	0.05	0.339
Lateral Anchor Shell	1.00	m	20.64	m²/m	20.64	m²	200	0.05	0.206
Lowering Padeye	1.10	m²	2.00	m²/m²	2.20	m²	200	0.05	0.022
Mooring Chain	18.00	m	1.37	m²/m	24.66	m²	200	0.05	0.247
Total					**81.41**				**0.814**
2. Surface in Mud									
Lateral Anchor Shell	20.00	m	20.64	m²/m	412.80	m²	20	1.00	8.256
Lowering Padeye	0.75	m²	2.00	m²/m²	1.50	m²	20	1.00	0.030
Moring Chain	12.00	m	1.37	m²/m	16.44	m²	20	1.00	0.329
Total					**430.74**				**8.615**
			Total Outside Surface		512.15	m²	Total Outside Surface Current		**9.429**

Table 8.5 Cathodic Protection Calculation Table at Average Stage Current Required for FPSO Mooring Pile

Component Description	Net Quantity	Unit	Surface to Protection (Per Unit Length or Surface)	Unit	Surface Protection	Unit	Current Density mA/m²	Coating Breakingdown Factor	Current Required A
OUTSIDE SURFACE									
1. Surface In Sea water									
Head Plate	33.91	m²	1.00	m²/m²	33.91	m²	100	0.36	1.221
Lateral Anchor Shell	1.00	m	20.64	m²/m	20.64	m²	100	0.36	0.743
Lowering Padeye	1.10	m²	2.00	m²/m²	2.20	m²	100	0.36	0.079
Mooring Chain	18.00	m	1.37	m²/m	24.66	m²	100	0.36	0.888
Total					**81.41**				**2.931**
2. Surface in Mud									
Lateral Anchor Shell	20.00	m	20.64	m²/m	412.80	m²	20	1.00	8.256
Lowering Padeye	0.75	m²	2.00	m²/m²	1.50	m²	20	1.00	0.030
Moring Chain	12.00	m	1.37	m²/m	16.44	m²	20	1.00	0.329
Total					**430.74**				**8.615**
			Total Outside Surface		512.15	m²		Total Outside Surface Current	11.546

MATTHEW OMOTOSO, PH.D

Table 8.6 Cathodic Protection Calculation Table at Final Stage Current Required for FPSO Mooring Pile

Component Description	Net Quantity	Unit	Surface to Protection (Per Unit Length or Surface)	Unit	Surface Protection	Unit	Current Density mA/m²	Coating Breakingdown Factor	Current Required A
OUTSIDE SURFACE									
1. Surface In Sea water									
Head Plate	33.91	m²	1.00	m²/m²	33.91	m²	130	0.68	2.998
Lateral Anchor Shell	1.00	m	20.64	m²/m	20.64	m²	130	0.68	1.825
Lowering Padeye	1.10	m²	2.00	m²/m²	2.20	m²	130	0.68	0.194
Mooring Chain	18.00	m	1.37	m²/m	24.66	m²	130	0.68	2.180
Total					**81.41**				**7.197**
2. Surface in Mud									
Lateral Anchor Shell	20.00	m	20.64	m²/m	412.80	m²	20	1.00	8.256
Lowering Padeye	0.75	m²	2.00	m²/m²	1.50	m²	20	1.00	0.030
Moring Chain	12.00	m	1.37	m²/m	16.44	m²	20	1.00	0.329
Total					**430.74**				**8.615**
			Total Outside Surface		**512.15**	m²		**Total Outside Surface Current**	**15.811**

Anode Number

Calculate the number of required anodes based on the total mass of anode using equation 8.11.

$$Q = \frac{Mt}{Unit \text{ weight}} \qquad \qquad \textbf{(8.11)}$$

$$Q = \frac{1581}{65} = 24.32 \text{ anodes} \Rightarrow (+ 10\%) = 27 \text{ anodes}$$

An increase of 10 percent in the number of anodes has been taken for anode quantity by mass.

Anode Quantity by Current

Calculate the number of required anodes by current at various stages.

$$Q = \frac{I \, required}{Output} \qquad \qquad \textbf{(8.12)}$$

At initial condition: $\frac{9.429}{1.678} = 5.62 \Rightarrow (+10\%) = 7$ anodes

At final condition: $\frac{15.81}{0.929} = 17.0$ anodes $\Rightarrow (+10\%) = 19$ anodes

Figure 8.7 Anode Quantity Summaries

Stage/Condition	Quantity by Current	Quantity by Weight
Initial	7	-
Average	19	-
Final	-	27

The number of anodes to ensure the protection of FPSO suction anchor pile during the design life is twenty-seven.

MATTHEW OMOTOSO, PH.D

Final Check

As a final check, verify that the anode current drawing capacity (I_d) is greater than the initial polarizing required current (I_p).

Current Drawing Capacity

Anode current drawing capacity is calculated using equation 8.13.

$$I_d = \frac{0.25 \times N}{R_0}$$

(8.13)

N is the number of anodes to ensure the protection of FPSO suction anchor piles during its design life (27 in number). R_0 is anode resistance at the initial stage that could be determined, using equation 8.14.

$$R_0 = \frac{0.315\rho}{\sqrt{A}}$$

(8.14)

$$R_0 = \frac{0.315 \times 30}{\sqrt{2(22 \times 18 + 56 \times 18) + 22 \times 56}} = 0.149$$

$$Id = \frac{0.25 \times 27}{0.149} = 45.30\,amps$$

$$Id \geq 1.1 \times Ip$$

Initial polarizing required current (I_p) = 9.43 amps

$I_d \geq 1.1 \times 9.43 = 10.37$ amps

45.30 amps > 10.37 amps

Conclusion

Therefore, 27 anodes of aluminium 560 × 220 × 180 × 65 kg are sufficient to protect the suction pile for the duration of twenty-five years.

8.2 Subsea Hardware Cathodic Protection

Example -2

Determine the cathodic protection system design for a subsea satellite well completion in 150-meter water depth in offshore Nigeria. The equipment area is 2,397 square meters, and the approximate area of 27.87 square meters is charged for 304 stainless steel parts. This is a single well completion, and the cathodic protection will be provided by aluminum sacrificial anodes. Only the outside surfaces of the equipment will be protected. The design life of the cathodic protection system is proposed for ten years. The coating breakdown factor is considered to be 25 percent.

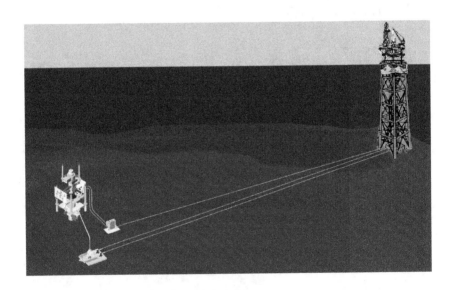

Figure 8.3 Typical single well completions

Surface Area Calculations

The following surfaces have been considered for the cathodic protection calculations, and detailed computations are shown in tables 8.9, 8.10, and 8.11.

Equipment surfaces of carbon steel: 222.69 m²
Equipment surfaces of stainless steel: 27.87 m²

Design Current Densities

The following current densities were considered for the submerged zone at a depth up to thirty meters in the temperate region.

Initial:	200 mA/m²
Mean:	100 mA/m²
Final:	130 mA/m²

For the mud zone, 20 mA/m² was adopted as the design current density.

Anodes Characteristics (Aluminum-Based Anodes)

Aluminium base anodes and anode shape and dimensions were chosen from economics point of view and design experience.

Cross section:	128 × 127 cm²
Total length:	1,000.0 cm
Weight:	38.8 kg
Density:	2.75 kg/dm

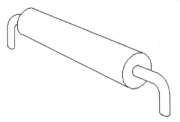

Figure 8.4 Anode diagram

Protection Current Densities

Here is the quantity of current required to protect a structure against corrosion, as provided in Table 8.8.

$$I_c = (A_c) \times (f_c) \times (i_c) \tag{8.15}$$

I_c is the current density (initial, average, final), A_c is the individual area, f_c represents coating breakdown factor, and i_c is the design current density.

Coating Breakdown Factor

Allow a coating breakdown factor of 25 percent for the painted parts of the subsea equipment.

Required Current Intensity

The required current intensity summary presented in table 8.8 is taken from tables 8.9, 8.10, and 8.11.

Determination of Anode Quantity

The total anode mass M_t required to maintain cathodic protection throughout the design life is given by equation 8.16.

$$M_t = D \times I_R \times \frac{m}{u} \qquad \text{(8.16)}$$

D = design life (10 years)

I_R = required intensity (average)

u = utilization factor; a value of 0.80 has been taken for all the anodes (section 7.4.8)

$$m = \text{consumption rate} = \frac{8760}{\varepsilon} = 4.38 \qquad \text{(8.17)}$$

ε = the electrochemical efficiency; a value of 2,000 Ah/kg for an aluminum-base anode is adopted.

Table 8.8 Required current intensity summaries

Protected Zone	Average	Initial Stage	Final Stage
External Surfaces (Submerged & Buried)	10.74 A	19.10 A	13.25 A

Total Mass of Anodes

$$M_t = 10 \times 10.74 \times \frac{4.38}{0.80} = 558.021 kg$$

Anode Number

Calculate the number of required anodes with equation 8.18.

$$Q = \frac{Mt}{\textit{Unit } \text{weight}} \tag{8.18}$$

$$Q = \frac{558.02}{38.8} = 14.38 \text{ anodes} \Rightarrow (+10\%) \ 16 \text{ anodes}$$

An increase of 10 percent in the number of anodes has been taken for provision of surfaces that have not been input.

Final Check

As a final check, verify that the anode current drawing capacity (I_d) is greater than the initial polarizing required current (I_p); R is anode resistance at the initial stage for long and standoff anode.

Current Drawing Capacity

$$I_d = \frac{0.25 \times N}{R} \tag{8.19}$$

$$R = \frac{\rho}{2\pi L}\left(In\frac{4L}{r} - 1\right) \tag{8.20}$$

where:

$$r = \sqrt{\frac{A}{\partial}} \tag{8.21}$$

L = 1m, A = 0.016 m², r = 0.71 m

$$R = \frac{\rho}{2\pi L}\left(In\frac{4L}{r} - 1\right) = 0.12 \text{ ohm} \tag{8.22}$$

MATTHEW OMOTOSO, PH.D

$$I_d = \frac{0.25 \times 16}{0.13} = 30.80 \text{ amps}$$

$$I_d = 1.1 \times I_p \qquad\qquad\qquad \textbf{(8.23)}$$

Initial polarizing required current (I_p) = 17.42 amps

$I_d \geq 1.1 \times 19.10 = 21.10$ amps

21.10 amp > 13.80 amp

Conclusion

Therefore 16 anodes of aluminium 1,000 × 129 × 128 × 38.8 kg are sufficient.

Table 8.9 Cathodic Protection Calculation Table at Initial Stage Current Required for Subsea Satellite Well Completion

Component Description	Net Quantity	Unit	Surface to Protection (Per Unit Length or Surface)	Unit	Surface Protection	Unit	Current Density mA/m²	Coating Breakingdown Factor	Current Required A
OUTSIDE SURFACE									
1. Surface In Sea water									
Carbon Steel Parts	222.69	m²	1.0	m²/m²	222.69	m²	200	0.25	11.135
Stainless Steel Part	27.87	m²	1.0	m²/m²	27.87	m²	200	1.00	5.574
Total					250.56				16.709
2. Surface in Mud									
0.782m Dia Conductor Pipe	50.00	m	2.39	m²/m	119.50	m²	20	1.00	2.390
Total					119.50				2.390
			Total Outside Surface	m²	370.06	Total Outside Surface Current			19.099

MATTHEW OMOTOSO, PH.D

Table 8.10 Cathodic Protection Calculation Table at Average Stage Current Required for Subsea Satellite Well Completion

Component Description	Net Quantity	Unit	Surface to Protection (Per Unit Length or Surface)	Unit	Surface Protection	Unit	Current Density mA/m²	Coating Breakingdown Factor	Current Required A
OUTSIDE SURFACE									
1. Surface In Sea water									
Carbon Steel Parts	222.69	m²	1.0	m²/m²	222.69	m²	100	0.25	5.567
Stainless Steel Part	27.87	m²	1.0	m²/m²	27.87	m²	100	1.00	2.787
Total					**250.56**				**8.354**
2. Surface in Mud									
0.782m Dia Conductor Pipe	50.00	m	2.39	m²/m	119.50	m²	20	1.00	2.390
Total					**119.50**				**2.390**
			Total Outside Surface	m²	370.06	**Total Outside Surface Current**			10.744

Table 8.11 Cathodic Protection Calculation Table at Final Stage Current Required for Subsea Satellite Well Completion

Component Description	Net Quantity	Unit	Surface to Protection (Per Unit Length or Surface)	Unit	Surface Protection	Unit	Current Density mA/m²	Coating Breakingdown Factor	Current Required A
OUTSIDE SURFACE									
1. Surface In Sea water									
Carbon Steel Parts	222.69	m²	1.0	m²/m²	222.69	m²	130	0.25	7.237
Stainless Steel Part	27.87	m²	1.0	m²/m²	27.87	m²	130	1.00	3.623
Total					250.56				10.861
2. Surface in Mud									
0.782m Dia Conductor Pipe	50.00	m	2.39	m²/m	119.50	m²	20	1.00	2.390
Total					119.50				2.390
			Total Outside Surface	m²	370.06	Total Outside Surface Current			13.251

MATTHEW OMOTOSO, PH.D

8.3 Jacket Structures Cathodic Protection

Example -3

The platform will be installed offshore Nigeria in a water depth of 97 meters. The platform consists of a piled steel jacket with four legs and four immersed plans at elevations -12.0 meters, -32.0 meters, -53.0 meters, and -77.0 meters, with respect to lowest astronomical tide (LAT). A fifth plan, at elevation +5.0 meters, is located in the emerged part of the jacket. The jacket is anchored to the seabed by four piles driven to a depth of 70 meters. For the structural elements in the splash zone (from +4.5 to -1.0 meters), a protective coating has been provided, and the immersed parts of the jacket are uncoated. The cathodic protection system shall be designed to guarantee full protection of the structure for 25 years.

Figure 8.5 Fixed jacket structures

Reference Documents

- Det Norske Veritas, RP B401, 1993, "Cathodic Protection Design"
- CEN EN 12473, January 2000, "General Principles of Cathodic Protection in Sea Water"
- CEN EN 12495, January 2000, "Cathodic Protection for Fixed Steel Offshore Structures"

Surface Area to Protect

To assess the local protection current demand, the structure has been divided into the following zones.

- Emerged zone, without cathodic protection:	above +4.5 m
- Splash zone:	from +4.5m to -1.0 m
- First trunk:	from -1.0m to -12.0 m
- Second trunk:	from -12.0 m to -32.0 m
- Third trunk:	from -32.0 m to -53.0 m
- Fourth trunk:	from -53.0 m to -77.0 m
- Fifth trunk:	from -77.0 m to -97.0 m
- Buried zone	

Each of the trunks includes the vertical structural elements between the relevant elevations and the horizontal elements in the lower plan of the trunk, if there are any. Also, in the immersed part of the jacket, the following steel elements have been considered.

- 12N0 conductor pipes, diameter 30" (762 mm)
- 2N0 J-tube diameter 18" (457.2 mm) and 8" (203.2 mm)
- 1N0 casing, up to -15 m, diameter 24" (609.6 mm)
- 1N0 sump, up to -35 m, diameter 1,300 mm

MATTHEW OMOTOSO, PH.D

- 1N0 CPM caisson, diameter 219 mm
- 1N0 cone end for each leg located on the seabed, higher diameter 4.0 m, lower diameter 1.8 m, and height 2.0 m
- clamps
- Four piles buried for 70 m into the seabed, diameter 72" (1,829 mm)

The surfaces in the buried zone include the fixing piles and well casings. The calculated surface area to be protected and separated into trunks, as described earlier, are reported shortly. To account for minor metal surfaces not explicitly considered in the calculations, the following safety factors have been adopted: surface area exposed to sea water, 1.15; surface area exposed to sea mud, 1.10.

Splash Zone

-	bare jacket:	244 m^2
-	conductor pipes:	158 m^2
-	J-tubes:	12 m^2
-	casing:	11 m^2
-	sump:	22 m^2
-	monitoring pipe:	4 m^2
-	clamps:	5 m^2
-	**Total**	**456 m^2**

First Trunk

-	bare jacket:	765 m^2
-	conductor pipes:	316 m^2
-	J-tubes:	23 m^2
-	casing:	21 m^2
-	sump:	45 m^2
-	monitoring pipe:	8 m^2
-	clamps:	11 m^2
-	**Total**	**1,188 m^2**

Second Trunk

-	bare jacket:	1,290 m²
-	conductor pipes:	574 m²
-	J-tubes:	42 m²
-	casing:	6 m²
-	sump:	82 m²
-	monitoring pipe:	14 m²
-	clamps:	16 m²
-	**Total**	**2,024 m²**

Third Trunk

-	bare jacket:	1,481 m²
-	conductor pipes:	603 m²
-	J-tubes:	45 m²
-	sump:	12 m²
-	monitoring pipe:	15 m²
-	clamps:	14 m²
-	**Total**	**2,170 m²**

Fourth Trunk

-	bare jacket:	2,155 m²
-	conductor pipes:	689 m²
-	J-tubes:	51 m²
-	monitoring pipe:	17 m²
-	clamps:	14 m²
-	**Total**	**2,926 m²**

Fifth Trunk

-	bare jacket:	896 m²
-	conductor pipes:	574 m²
-	J-tubes:	41 m²
-	monitoring pipe:	14 m²
-	cones end:	83 m²

- clamps:	14 m²	
- **Total**	**1,622 m²**	

Buried (Sea Mud) Zone

- piles:	1,608 m²	
- **Total**	**1,608 m²**	

Summary

- Total surface area relevant to the splash zone:
- coated surface: **456 m²**
- Total surface area immersed to sea water:
- uncoated surface: **9931 m²**
- Total surface area exposed to sea mud:
- uncoated surface: **1,608 m²**

Table 8.12 Current density values

Zone to Be Protected	Current Density (mA/m²)		
	Initial	Average	Final
Immersed zone (seawater)	130	70	90
Marine sediments (sea mud)	25	20	20

An anode (aluminum-zinc-indium alloy sacrificial anodes, slender type) with the following parameters has been selected to use in the design.

- length:	2,337	mm
- width:	226/206	mm
- height:	190	mm
- steel insert:	4"	Sch. 80
- net aluminum mass:	217	kg
- gross weight:	283	Kg

- capacity (ε):   2,500   Ah/kg
- utilization factor: 90%

Anode Resistance

The anode resistance (R_a) is given by the Dwight formula.

$$R = \frac{\rho}{2\pi L}\left(In\frac{4L}{r} - 1\right) \tag{8.24}$$

$$r = \sqrt{\frac{A}{\pi}} \tag{8.25}$$

where ρ is environmental resistivity and A is exposed anode surface area.
- Initial resistance -0.06 Ω
- Initial resistance -0.06 Ω

Resistivity

The following value has been assumed for water resistivity.
- Seawater: 30-ohm cm

Anode Current Output

$$I = \frac{\Delta E}{R} \tag{8.26}$$

$$\Delta E = E_c - E_a \tag{8.27}$$

E_c = protective potential = -0.800 V

E_a = alloy potential = -1.05 V

At initial condition: $= \dfrac{0.25}{0.06} = 4.12$ A

At final condition: $= \dfrac{0.25}{0.07} = 3.57$ A

Coating Breakdown Factor

$$f_{cm} = a + \frac{b \times t_f}{2} \qquad\qquad (8.28)$$

$$f_{cf} = a + b \times t_f \qquad\qquad (8.29)$$

Initial $f_c = a$; t_f is the design life

For category II coating, $a = 0.05$ and $b = 0.025$.

Table 8.13 Coating breakdown factor

S/N	Stages	Coating Breakdown Factor
1	Initial	5%
2	Average/Medium	36%
3	Final	68%

Table 8.14 Current Required for the Jacket Structure Corrosion Protection

S/N	COMPONENT DESCRIPTION	SURFACE TO PROTECT	UNIT	CURRENT DENSITY mA/m^2	COATING BREAKDAWN FACTOR	CURRENT REQUIRED A
	INITIAL STAGE					
	Surface in Seawater					
1	Splash Zone	456	m^2	130	0.05	2.96
2	First Truck	1188	m^2	130	1.00	154.44
3	Second Truck	2024	m^2	130	1.00	263.12
4	Third Truck	2170	m^2	130	1.00	282.1
5	Fourth Truck	2926	m^2	130	1.00	380.38
6	Fifth Trunk	1622	m^2	130	1.00	210.86
7	Buried Zone	1602	m^2	130	1.00	208.26
	AVERAGE STAGE					
	Surface in Seawater					
1	Splash Zone	456	m^2	70	0.36	11.49
2	First Truck	1188	m^2	70	1.00	83.16
3	Second Truck	2024	m^2	70	1.00	141.68
4	Third Truck	2170	m^2	70	1.00	151.9
5	Fourth Truck	2926	m^2	70	1.00	204.82
6	Fifth Trunk	1622	m^2	70	1.00	113.54
7	Buried Zone	1602	m^2	70	1.00	112.14
	FINAL STAGE					
	Surface in Seawater					
1	Splash Zone	456	m^2	90	0.68	27.91
2	First Truck	1188	m^2	90	1.00	106.92
3	Second Truck	2024	m^2	90	1.00	182.16
4	Third Truck	2170	m^2	90	1.00	195.3
5	Fourth Truck	2926	m^2	90	1.00	263.34
6	Fifth Trunk	1622	m^2	90	1.00	145.98
7	Buried Zone	1602	m^2	90	1.00	144.18

Protection Current Densities

The quantity of current required to protect a structure against corrosion is provided in table 8.14.

$$I_c = \left(A_c \right) \times \left(f_c \right) \times \left(i_c \right) \tag{8.30}$$

Here, I_c is current density (initial, average, final), A_c is individual area, f_c represents coating breakdown factor, and i_c is design current density.

The total protection current required for individual components of the jacket structures is calculated and presented in table 8.12 based on the current density provided in table 8.11. The coating breakdown factor was considered only for the splash zone.

Protection Potential

Cathodic protection conditions are expressed by the following limits.

- positive limit: -0.80 V vs. Ag/AgCl/seawater
- negative limit: -1.05 V vs. Ag/AgCl/seawater

Determination of Anode Quantity

The total anode mass (M_t) required for maintaining cathodic protection throughout the design life (D) is given by the equation below, where u is utilization factor, I_R is required current; and ε is the electrochemical efficiency.

$$M_t = D \times I_R \times \frac{m}{u} \qquad \text{(8.31)}$$

$$m = \text{consumption rate} = \frac{8760}{\varepsilon} = 3.50 \qquad \text{(8.32)}$$

Anode Quantity

Q is the quantity of required anode to protect the system.

$$Q = \frac{Mt}{Unit\ weight} \qquad\qquad (8.33)$$

Anode Quantity by Current

Calculate the number of required anodes by current at various stages.

$$Q_I = \frac{I_I\ required}{Output} \qquad \text{—Initial Condition} \qquad (8.34)$$

$$Q_F = \frac{I_F\ required}{Output} \qquad \text{—Final Condition} \qquad (8.35)$$

Anode Quantities Summary

The calculated number of anodes required to protect the structures based on initial, average, and final conditions are present in table 8.15.

Table 8.15 Anode quantities summary

Component Description	Initial	Average	Final
First trunk	69	37	48
Second trunk	118	63	82
Third trunk	126	68	88
Fourth trunk	170	92	118
Fifth trunk	94	51	65
Total	577	311	401

The minimum anode number for each trunk is as follows.

First trunk	69
Second trunk	118

Third trunk	126
Fourth trunk:	170
Fifth trunk:	94
Total minimum anode requirements:	**577**

The piles surfaces exposed to sea mud are protected by anodes exposed to the seawater and located on the lower part of the steel structure. The number of the anodes required for the pile protection is about 93 based on a separate calculation that is not present in this edition.

Table 8.16 Calculation of anode quantity by mass and required current

S/N	COMPONENT DESCRIPTION	CURRENT REQUIRED A	DESIGN LIFE (Year)	COMSUMPTION RATE	UTILIZATION FACTOR	TOTAL ANODE MASS (KG)	ANODE UNIT WEIGHT (KG)	ANODE QUANTITY BY MASS	CURRENT OUTPUT (A)	ANODE QUANTITY BY CURRENT
	INITIAL STAGE									
	Surface in Seawater									
1	Splash Zone	2.96	25	3.5	0.9	288	217	1	4.12	1
2	First Truck	154.44	25	3.5	0.9	15,015	217	69	4.12	37
3	Second Truck	263.12	25	3.5	0.9	25,581	217	118	4.12	64
4	Third Truck	282.1	25	3.5	0.9	27,426	217	126	4.12	68
5	Fourth Truck	380.38	25	3.5	0.9	36,981	217	170	4.12	92
6	Fifth Trunk	210.86	25	3.5	0.9	20,500	218	94	5.12	41
7	Buried Zone	208.26	25	3.5	0.9	20,248	217	93	4.12	51
	AVERAGE STAGE									
	Surface in Seawater									
1	Splash Zone	11.5	25	3.5	0.9	1,117	217	5	N/A	N/A
2	First Truck	83.16	25	3.5	0.9	8,085	217	37	N/A	N/A
3	Second Truck	141.68	25	3.5	0.9	13,774	217	63	N/A	N/A
4	Third Truck	151.9	25	3.5	0.9	14,768	217	68	N/A	N/A
5	Fourth Truck	204.82	25	3.5	0.9	19,913	217	92	N/A	N/A
6	Fifth Trunk	113.54	25	3.5	0.9	11,039	217	51	N/A	N/A
7	Buried Zone	112.14	25	3.5	0.9	10,903	217	50	N/A	N/A
	FINAL STAGE									
	Surface in Seawater									
1	Splash Zone	27.9	25	3.5	0.9	2,713	217	13	3.57	8
2	First Truck	106.92	25	3.5	0.9	10,395	217	48	3.57	30
3	Second Truck	182.16	25	3.5	0.9	17,710	217	82	3.57	51
4	Third Truck	195.3	25	3.5	0.9	18,988	217	88	3.57	55
5	Fourth Truck	263.34	25	3.5	0.9	25,603	217	118	3.57	74
6	Fifth Trunk	145.98	25	3.5	0.9	14,193	218	65	4.57	32
7	Buried Zone	144.18	25	3.5	0.9	14,018	217	65	3.57	40

Final Check

As a final check, verify that the anode current drawing capacity (I_d) is greater than the initial polarizing required current (I_p). Anode current drawing capacity is calculated based on the equation below.

$$I_d = \frac{0.25 \times N}{R} \tag{8.36}$$

N is the number of anodes to ensure the protection of the jacket structures during its design life (577 in number). R is anode resistance at the initial stage.

$$R = \frac{\rho}{2\pi L}\left(In\frac{4L}{r} - 1 \right) = 0.06 \text{ ohm} \tag{8.37}$$

$$Id = \frac{0.25 \times 577}{0.06} = 302,404.17 \text{ amps}$$

$$Id = 1.1 \times Ip \tag{8.38}$$

Initial polarizing required current (I_p) = 9,930 amps

$I_d \geq 1.1 \times 9930 = 10,923$ amps

302,404.17 > 9,930 amps

Therefore, 577 anodes of aluminium 2,337 × 226 × 190 × 217 kg are sufficient to protect the jacket structures for a period of 25 years.

Monitoring System Design

The cathodic protection monitoring system shall be in accordance with Det Norske Veritas, RP B401, for the offshore platform cathodic protection monitoring system with sacrificial anodes. The monitoring system shall consist of the following.

- Proper number of monitored anodes
- Data acquisition unit
- Underwater cables
- Proper number of reference cells to measure the steel potential

The reference cells shall be silver or silver chloride (Ag/AgCl/seawater) placed near jacket complex nodes and beams.

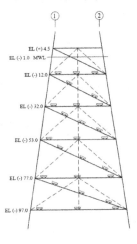

Figure 8.6 Jacket structure anode arrangements

8.4 Crude Oil Pipeline Cathodic Protection

Example -4

Determine the cathodic protection system for a well-coated crude oil pipeline of 6 inches (150 millimeter) in diameter and 18 kilometers long, running from a production wellhead on land fields to a flow station. Application of impressed current cathodic protection system is recommended using

MATTHEW OMOTOSO, PH.D

circular, high-silicon, chromium-bearing, cast-iron anodes. The life span of the facility is proposed to be twenty years.

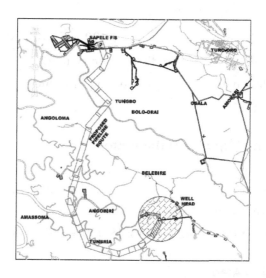

Figure 8.7 Pipeline route showing in Google

Table 8.17 Design data

Descriptions	Values
Pipe Diameter	6 in (150mm)
Average Soil Resistivity (ρs)	2,000 ohms-centimeters
Pipeline Life span	20 years
Current Density (ρ)	30 mA/m^2
Coating Efficiency (f)	50%
Coating Resistance (R)	681 ohms/m^2
Maximum Groundbed Resistance	2 ohms

- The pipeline must be isolated from the flow station and wellhead with an insulating joint.

- High-silicon chromium-bearing cast iron anodes must be used with carbonaceous backfill for the anode.
- The pipe is coated with hot-applied coal-tar enamel holiday-checked before installation.
- Electricity power is available at 120/240 volts AC single phase from a nearby overhead distribution system.
- The design for 50 percent coating efficiency is based on past experience.

Determination of Pipeline Surface Area

Pipe size: 0.15 m

Pipe length: 1,8000 m

Pipe area: $A_p = \pi dL = 3.142 \times 0.15 \times 18,000 = 8,483.4 \text{ m}^2$ **(8.39)**

Total Protection Area

$A_c = f \times A_p = 50\% \times 8,483.4 = 4,241.7 \text{ m}^2$ **(8.40)**

Total Protection Area

$I_P = \rho \times A_c = 30 \times 4,241.7 = 12,7250 \text{ mA} = 127.25 \text{ A}$ **(8.41)**

Anode Selection

An HSCBCI anode with the following characteristics was chosen.

Table 8.18 Anode data

Descriptions	Values
Material	Iron Silicon Chromium
Anode Spacing	66 ft (20 m)
Anode Shape Factor (K)	0.0167

Cross Section Area	4 ft² (0.371 m²)
Weight	110 lb (50 kg)
Diameter	10 in (2.54 cm)
Length	84 in (213.4 cm)

Anode Quantity

i. Calculate the number of anodes needed to meet the anode supplier's current density limitations. The recommended maximum current density output for HSCBCI anode is 100 A/m²

$$N = \frac{I_p}{A_1 \times I_1} = \frac{127.25}{0.371 \times 100} = 3.4 \approx 4 \text{ anodes} \qquad (8.42)$$

ii. Calculate the number of anodes required to meet the design life requirements.

$$N = \frac{L_f \times I}{1000W} = \frac{20 \times 127250}{1000 \times 110} = 23.13 \approx 24 \text{ anodes} \qquad (8.43)$$

iii. Calculate the anode quantity required to meet the minimum anode groundbed resistance requirement, where R_a is groundbed resistance; S is anode spacing (66 ft), L is anode back fill column (7 ft), and P is the parallel factor (-0.00121; see table 7.3)

$$N = \frac{\rho_s \times K}{L\left(R_a - \dfrac{\rho_s}{S \times P}\right)} \qquad (8.44)$$

$$N = \frac{2000 \times 0.0167}{7\left(0.191 - \dfrac{2000}{20 \times 0.00121}\right)} = \approx 1 \text{anode}$$

Anode Groundbed Resistance

$$R_a = \frac{\rho_S \times K}{N \times L} + \frac{\rho \times P}{S} \qquad (8.45)$$

$$R_a = \frac{2000 \times 0.0167}{24 \times 7} + \frac{2000 \times 0.0121}{66} = 0.564 \text{ ohms}$$

Select Number of Anode

Because the second calculation resulted in the largest number of anodes, it will be considered; therefore, thirty-four anodes will be used.

Groundbed Resistance for Header Cable

The estimated length of cable to be used is 130 meters. The resistance specified by the manufacturer is 0.00481 ohm per 30 meters of No. 2 AWG cable.

$$R_W = \left(\frac{ohms}{R_L} \right) \times L \qquad (8.46)$$

$$R_W = \left(\frac{0.00480}{30} \right) \times 130 = 0.021 \text{ ohms}$$

Structure-to-Electrolyte Resistance

The structure-to-electrolyte resistance can be calculated using the subsequent equation, where, A_{co} is the coated pipe area.

$$R_C = \frac{R}{A_{co}} = \frac{681}{8486} = 0.080 \text{ ohms} \qquad (8.47)$$

$$A_{Co} = \pi dL = 3.142 \text{x } 0.15 \text{ x } 18000 = 8483.4 \text{m}^2$$

Total Resistance

$$R_T = R_a + R_W + R_C \qquad (8.48)$$

0.564 ohms + 0.021 ohms + 0.080 ohms = 0.665 ohms

Rectifier Voltage

$$V_{rec} = I_p \times R_T \times 150\% \qquad (8.49)$$

$$V_{rec} = 127.25 \times 0.665 \times 1.5 = 126.93V$$

$$P = (V_{rec}) \times (I) \qquad (8.50)$$

$$P = 126.93 \times 127.25 = 16,152W$$

Select the rectifier based on the design requirement of 126.93 volts, 127.25 amperes, and 16,152 watts.

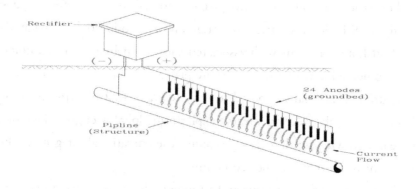

Figure 8.8 Anode groundbed layout

CHAPTER 9

FIXED OFFSHORE PLATFORM ASSESSMENT

9.0 Introduction

Offshore fixed platform is defined as a man-made "island" built to allow offshore crude oil production through conventional above-water techniques; see figure 9.1. Several barrels of crude oil are produced worldwide via fixed offshore platforms with associated subsea pipelines and manifolds. The submerged part of a fixed offshore platform is known as jacket. This structure is continually exposed to salty seawater that hastens corrosion damages and subsequently affects the platform global strength. Therefore, there is a growing need to closely monitor the operational integrity of the platforms to prevent unexpected failures.

The actions of ocean wave and strong winds against a fixed platform also lead to the development of fatigue cracks on the joints. Jacket structures may have been deteriorated to an undisclosed degree through decades of existence in deep seawater. Adequate safety of the structures can be achieved through assessment and appropriate revamp works. Therefore, corrosion and fatigue risks demand detailed study and investigation regarding how the hazard affects the fixed offshore platform with special reference to jacket structures.

The safety of fixed offshore platform is generally assumed to be achieved by design according to the established standards and procedures in order to prevent a catastrophic collapse that may be caused by component deterioration and associated risks. But there is a general recognition across the construction industry that the assessment method for existing structures is quite different from the new design process. The compliance with existing rules and regulations may contribute to the safety of a fixed offshore platform's safety in the design stage, but it may not be appropriate for the assessment of aging and corroded jacket structures.

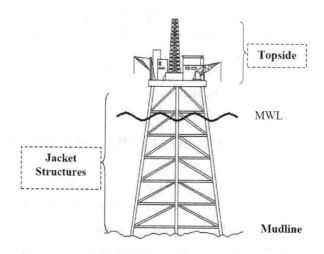

Figure 9.1: Schematic of offshore jacket platform

Large numbers of existing fixed offshore platforms for crude oil production in many parts of world have been designed for a life span of twenty-five years, as specified by API RP 2A WSD. The exorbitant cost of the structure replacement has let the majority of the operators exploit the platforms beyond the design life. The approximate age distribution for fixed offshore platform installations in the Niger Delta shows that relatively large number of the platform age is greater than twenty-five years, as shown in figure 9.2.

The likelihood of failure with time in service for civil engineering structures increases during the late life, which is often associated with degradation of construction materials and onset time-dependent failure mechanisms. Fixed offshore platform structure safety against environmental loads and operation stresses that was acceptable in the design life, may not be appropriate for the same structure later in its lifetime.

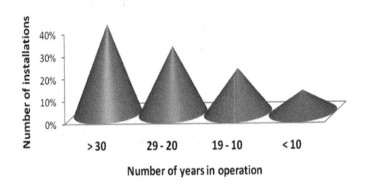

Figure 9.2 Fixed platforms age distribution in Niger Delta

The steel structures located in an offshore environment are subjected to corrosion degradation due to oxygenated and salty seawater. The section of an unprotected specimen generally has the highest corrosion rates around the splash zones. Therefore, legs, framing members, and boat landing areas of jacket platform structures suffer more corrosion damage than any other parts of the structures. However, the available basic design methods in the current standards used for new structures design may not be suitable for the assessment of existing platform structures with regard to corrosion damages, and regular structural reliability evaluation of the platform is significant to prevent the structures from sudden collapse. The above phenomena formed the foundation for continued research work in this field of engineering.

Material degradation due to corrosion is an issue of high importance, not least for fixed offshore platforms, pipelines, and components of other crude oil facilities process systems in general. The reliable operation of such systems is a key factor for the safety of personnel working in these facilities and has a major influence on the economic benefit to the operators. The classification societies and petroleum production companies regularly carried out inspection on these facilities for corrosion losses estimation using ultrasonic techniques; it is carried out as part of the continued certification of fixed offshore platform for safety. However, such data are rarely published and are kept in the operator archive. This information is usually considered to be confidential and not readily accessible because the facilities are still in operation.

9.1 Assessment Procedures in Accordance with API Standard

The aim of structural inspection and assessment is to ensure safety of the facilities while the structural element that is not meeting the evaluation criteria is strengthened. The API RP 2A WSD is the most frequently applied standard for the structural design and assessment of fixed offshore platforms in many parts of the world. The standard is of interest because it is the most accessible document that takes assessment of existing jacket structures to a detailed level. However, the standard affirmed is applicable only for the assessment of fixed offshore platforms designed in accordance with twentieth or earlier editions of the same API standard. Structures designed after the twenty-first edition should be assessed in accordance with the criteria originally used for the design. By this clause, API 2000 cannot be used for assessment of all existing fixed offshore platforms because some of the platforms were built before the establishment of the standard. Therefore, this assertion is considered to be one of API 200's limitations.

There are two possible analysis checks mentioned in API RP 2A WSD, design level analysis and ultimate strength analysis. The design level analysis procedures for assessment is similar to those used for new platform structural design in the area of safety factors application. However, lateral environmental load can be reduced to 85 percent of the hundred-year condition for the high consequence platforms, and to 50 percent for low-consequence platforms. The previously mentioned review document is a relevant standard for fixed platform structural assessment. Nevertheless, the standard did not adequately cover platform structural failure probability during the operating life cycle. Also, the assessment guidelines in the standard are based only on life safety and failure consequences, with little consideration for structural reliability and risk-based assessment of domineering offshore hazard. With regard to the existing fixed platform in this section, the focus will be on the structural safety during its operating lifetime. The issue will be whether the safety established in the initial design stage is still appropriate when the platform is still in service.

9.2 Existing Structures and New Structures Data

Assessment of existing structures is an important topic for experts working in the oil and gas industry, where rehabilitation, including repairs and upgrading of crude oil production facilities, represents a significant part of all construction activities. This is due to several circumstances that include the existing structures representing substantial, continually increasing economic contributions for the stakeholders. Hence, the users are interested in a new way of exploiting existing structures. Also, many existing structures do not fulfill requirements of currently valid standards, and the standard for the assessment and retrofitting of existing structures has not been fully developed.

Assessment of existing structures often requires knowledge overlapping the framework of standards for the design of new structures. However, there are significant differences in the data about the existing structures and the structures in the design stage, as recorded in figure 9.3. These details are therefore essential to be accounted for while carrying out structural assessment of existing facilities. For instance, the model of a new structure might have topology and dimensions altered during the engineering design works by the asset managers and the design parameters will not be fixed until the detail engineering design is completed. But the existing structures have dimensions fixed, and changing the structural member sizes and design parameters is forestalled.

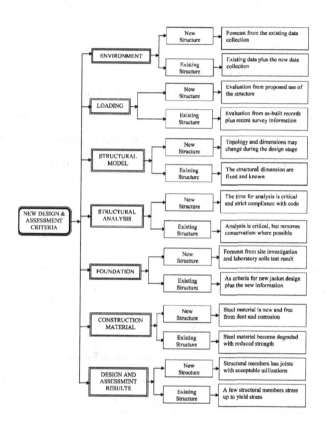

Figure 9.3: New structural designs versus structural assessments

For the existing structures, the incremental cost for increasing structural safety is usually high. For the new structures, the incremental cost of increasing structural safety is normally lower. In case of adequately managed structures, the available data are sufficient to enhance the structural analysis accuracy during assessment. But structures with little accessible data might have high uncertainty and lower structure safety margin. Further reading on this topic can be found in Kallaby et al. (1994) and Moan and Vardal (2001).

9.3 Existing Structures Assessment Hierarchy

There are set identify activities to be performed in order for validating the reliability of existing structures for continuous use, as represented in figure 9.4. These activities include investigation, structural analysis, verification, upgrading, monitoring, and probably a change in facility use. The information, through on-site and non-destructive examination, will be collected to establish the present condition of the structural members. Based on the site information and other documents, the structural analysis will be carried out to establish the reliability status followed by the required repair and upgrade.

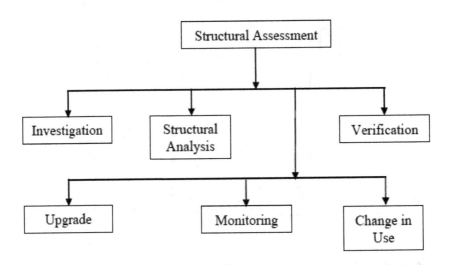

Figure 9.4: Structural assessment flow chart

9.4 Marine Structures Hazard Assessment

Fixed platform structure integrity assurance makes sure that the structures meet the required design purpose with great consideration for safety and reliability. It is a multidisciplinary activity that comes together for this purpose that includes inspection, material science, welding technology, structural analysis, and engineering safety. Hazard identification is significant for fixed platform structural failure prevention, and the method can be divided into three major components: hazard identification, assessment, and the maintenance program.

The key element of structural integrity assessment with regard to civil engineering structures is presented in figure 9.5. Diagnosing the structure hazard provides understanding of mechanisms that lead to structure deterioration and potential failure. Diagnosis enables design engineers to arrive at appropriate structural assessment methods. The structural hazard identification span from the inspection works on-site, such as using ultrasonic tester (UT), assessment activities off-site, and hazard resolution.

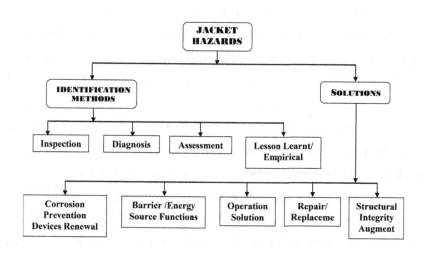

Figure 9.5 Structural reliability assessment key elements

The method used for determining hazard potential is dependent on the hazard characteristics, frequency, and the root causes. Because the design method used for new design may not be appropriate for the existing structure assessment, the basic criteria for choosing structure inspection and the assessment method depend on the technical capability to detect component defects and structural strength.

9.5 Fixed Platform Corrosion Damage

Facilities built in marine environments are prone to corrosion damage due to corrosive environment. Structures' corrosion damage may be very complex and may have limited information about the level of the damages. This is a particular concern where corrosion is not uniform along the depth of seawater, as may be applicable to several marine structures. It has been established from several site inspection results that the section of unprotected specimen located in the splash zones has the highest corrosion damage. Ultrasonic devices or the pulse-echo technique method is predominantly used to conduct surveillance on components to detect corrosion flaws and determine existing member thickness.

The device uses ultrasonic energy generated by transducers that change high-energy frequency signal into high-frequency mechanical energy. A liquid couplant to the metal wall transmits the sound wave generated by a transducer, and ultrasonic sound travels through the wall. The echo's sound waves are received and transformed into electrical impulses by the transducer. The device measures the time between the impulse and reflection, and the member current thickness can be calibrated, read, and recorded on a data sheet.

For a jacket structural member with long length, the UT test is performed on three spots along the member length. The spot with the minimum thickness is adopted as the current thickness. The fixed platform used as a case study was built twenty-three years ago in the Niger Delta. The nondestructive ultrasonic tester (UT) equipment was employed to establish the member thickness, and the results are presented in table 9.1. The platform sections and elevations are illustrated in figure 9.6. The data revealed that the member corrosion losses range from 0 percent to 17 percent, and the average corrosion loss was 4.5 percent of the original member thickness.

Table 9.1: Jacket Member Wall Thickness Corrosion Losses

Position	Member ID	Member Location and Elevation	Thickness (mm) 1985	UT Thickness (mm) 2008	Thickness Reduction (%) 2008
Row A	1DA	Horizontal Bracing EL (-) 1.5m	9.525	9.501	0.262
	3DA	Horizontal Bracing EL (-) 1.5m	9.525	9.45	0.787
	5DA	Diagonal Brace EL (-) 4.0m to (-) 7.0m	9.525	8.1	14.961
	5BA	Diagonal Brace EL (-) 7.0m to (-) 18.3m	12.7	10.5	17.323
	5AA	Diagonal Brace EL (-) 18.3m to (-) 32.0m	12.7	12.40	2.362
	5CA	Diagonal Brace EL (-) 18.3m to (-) 32.0m	12.7	11.90	6.299
	1DA	Diagonal Brace EL (-) 18.3m to (-) 32.0m	12.7	12.4	2.36
	3DA	Diagonal Brace EL (-) 18.3m to (-) 32.0m	12.7	11.9	6.30
Row B	1DB	Diagonal Brace EL (-) 4.0m to (-) 7.0m	9.525	9.6*	-
	3DB	Diagonal Brace EL (-) 7.0m to (-) 18.3m	12.7	12.7	0.00
	3MB	Diagonal Brace EL (-) 7.0m to (-) 18.3m	9.525	9.6*	-
	53B	Diagonal Brace EL (-) 18.3m to (-) 32.0m	12.7	11.6	8.66
	52B	Diagonal Brace EL (-) 18.3m to (-) 32.0m	12.7	11.9	6.30
	51B	Diagonal Brace EL (-) 18.3m to (-) 32.0m	12.7	12.5	1.575
Row 1	1D1	Diagonal Brace EL (-) 4.0m to (-) 7.0m	9.525	9.4	1.31
	3D1	Diagonal Brace EL (-) 7.0m to (-) 18.3m	9.525	9.1	4.46
Row 2	1D2	Diagonal Brace EL (-) 4.0m to (-) 7.0m	9.525	9.1	4.46
	3D2	Diagonal Brace EL (-) 18.3m to (-) 32.0m	9.525	8.9	6.56
	5B2	Diagonal Brace EL (-) 18.3m to (-) 32.0m	12.7	12.5	1.57
Plan @ (-) 7.0m	2MB	Horizontal Brace EL (-) 7.0m	9.525	9.2	3.412
	2MD	Diagonal Member EL (-) 7.0m	9.525	9.3	2.36
	2M2	Horizontal Brace EL (-) 7.0m	9.525	9.3	2.36
	2MH	Diagonal Brace EL (-) 7.0m	9.525	9.3	2.36
	2ME	Diagonal Brace EL (-) 7.0m	9.525	9.2	3.41
	2MA	Horizontal Brace EL (-) 7.0m	9.525	9.2	3.41
	2MF	Diagonal Brace EL (-) 7.0m	9.525	9.0	5.51
	2M1	Horizontal Brace EL (-) 7.0m	9.525	9.3	2.36
	2MG	Horizontal Brace EL (-) 7.0m	9.271	9.1	1.84
Plan @ (-) 18.3m	4MD	Diagonal Brace EL (-) 18.3m	9.525	9.1	4.46
	4ME	Diagonal Brace EL (-) 18.3m	9.525	9.0	5.512
	4MC	Diagonal Brace EL (-) 18.3m	9.525	9.1	4.46
	4MG	Diagonal Brace EL (-) 18.3m	9.525	8.6	9.71
	4M2	Horizontal Brace EL (-) 18.3m	9.525	9.1	5.512
	4MA	Horizontal Brace EL (-) 18.3m	9.525	9.0	9.711
	4M1	Horizontal Brace EL (-) 18.3m	9.525	8.6	4.462
	4MB	Horizontal Brace EL (-) 18.3m	9.525	9.1	1.844
	4MF	Horizontal Brace EL (-) 18.3m	9.271	9.1	1.84
Plan @ (-) 32.0m	6M2	Horizontal Brace EL (-) 32.0m	9.525	8.400	11.811
	6MA	Horizontal Brace EL (-) 32.0m	12.700	12.300	3.150
	6MD	Horizontal Brace EL (-) 32.0m	9.525	8.900	6.562
	6MC	Horizontal Brace EL (-) 32.0m	9.525	9.1	4.462
	6ME	Horizontal Brace EL (-) 32.0m	9.525	9.0	5.512
	6M1	Horizontal Brace EL (-) 32.0m	9.525	9.4	1.312
	6MB	Horizontal Brace EL (-) 32.0m	12.700	12.300	3.150
Jacket Legs	4MD	Jacket Leg – 1	19.1	18.11	5.18
	4ME	Jacket Leg – 2	19.1	18.25	4.45
	4MC	Jacket Leg – 3	19.1	18.52	3.04
	4MG	Jacket Leg – 4	19.1	18.65	2.36

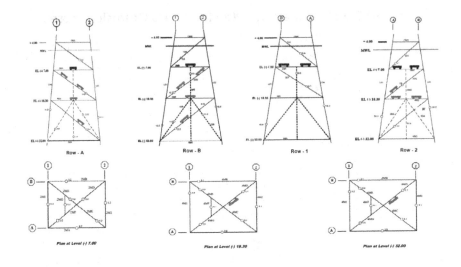

Figure 9.6 Jacket platform elevations and sections

9.6 Jacket Structure Fatigue Life Profile

This section is concerned with corrosion damage and investigation of joint fatigue life reduction trend because it was applicable to the fixed platform under the case study. The platform was modeled using SACS with due considerations to material deformation and nonlinearity in geometry. However, the analysis did not account for dynamic loading that is peculiar to slender deepwater platforms such as compliance structures because this study is largely an evaluation of joint fatigue life parameter, and this simplification was found satisfactory.

The computer model was based on the use of the SACS suite of software, version 5.2, from Engineering Dynamic Inc. A SACS model was built for the platform using existing engineering drawings and current member thickness. The fixed platform model is represented correctly by member cross-sectional properties, joint eccentricities, and end fixities. The 3-D structural model consists of an integrated model of substructures (jacket) and superstructure (topsides), having a detailed representation of primary members, risers, and conductors, together with a simplified representation of the elements causing hydrodynamic effects, such as anodes, marine growth, boat bumpers, and other appurtenances; see figure 9.7.

The riser model and pile-soil interaction were developed and incorporate into SACS model. The platform was designed for the combinations of dead load, live load, riser loads, and environmental loads. The marine growth profile used for the design was, as per the environmental data collected, during the site inspection. The platform legs and the other structural members that were subject to corrosion were modeled as structural members with reduced diameter and thicknesses to reflect the decayed states. The fatigue life damage and corresponding fatigue lives are generated using the fatigue modules. The member force spread transfer functions, response spectra and damage are computed.

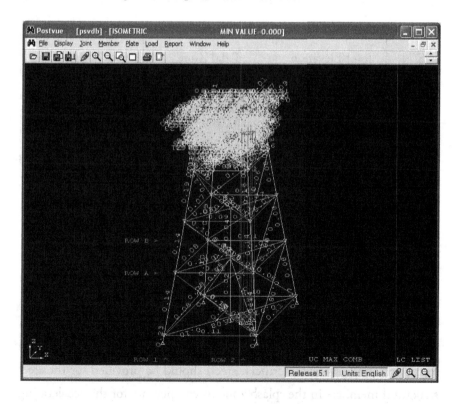

Figure 9.7 Fixed platform SACS model

The fatigue life of the jacket structure joints was reviewed to capture the number of joints with fatigue life less than ninety years, as expected for the jacket structure with a life span of thirty years with the safety factor of 3.0. The jacket structure was divided into four distinct tidal zones.

(1) Air Space Zone (above splash zone)
(2) Splash Zone
(3) Medium tide zone
(4) Low tide zone

Joints with fatigue life less than ninety years located at each of the tidal zones were recorded and presented in table 9.2.

Table 9.2 Joint fatigue life

S/N	Joint Elevation	Tidal Zone	Quantity of Joint with Fatigue Life Less Than 90 Years
1	Above +3	Air Zone	7
2	+3 to -4	Splash Zone	18
3	-4 to -18.3	Medium Tide Zone	11
4	-18.3 to -32	Low Tide Zone	8

The data in table 9.2 is represented with a diagram in figure 9.8, which is proposed in this textbook to be known as the jacket fatigue life profile. The profile indicates the number of joints with fatigue life less than ninety years, which was due to the function of corrosion rate and damage in each tidal zone. The diagram shows that there were more joints with less fatigue life in the splash zone than any other tidal zones. This diagram further supports the mitigation measure that additional thickness of steel for corrosion allowance (sacrificial steel) should be provided for the steel structural members in the splash zone to compensate for the accelerating corrosion losses.

A known corrosion allowance is of the order of 3–12 millimeters depending on design specifications and applicable codes. However, a minimum of 3 millimeters is recommended to the structural members seated in the splash zone for a well-protected structure. A well-protected structure means structures incorporated with the appropriate design and maintained cathodic protection systems.

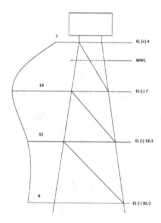

Figure 9.8 Jacket structure fatigue life profile

The outcome of this study offers a unique opportunity to determine the effectiveness of existing structural design standards and, if required, to develop recommendations for changes. The study provided an opportunity to evaluate the available design process for marine structures with regard to the relationship between corrosion and fatigue damage.

9.7 System Reliability Theoretical Background

The objective of this section is to present structures reliability assessment technique with regards to member corrosion losses. Marine structures members have been noted for time varying thickness reduction due to corrosion wastage that often lead to reduction in the structural global strength. Corrosion protection devices, such as barrier coatings and cathodic protection, are typically applied on marine installations. Despite the application of the protection devices, corrosion process is barely mitigated and not totally prevented.

Marine environments are good corrosive agents for mild and low-alloy steel structures, and for economic reasons, steels remained the preferred materials for engineering structures in a marine environment. The decline in marine structures strength is mostly due to corrosion and fatigue as the structures age and progressively corrode. With regard to fatigue damage, a fracture mechanics analysis is performed to establish the average time for a postulated initial crack to grow through the plate's thickness. When cracks occur in a nonredundant component, it may lead to structural failure, and therefore target reliability should be set higher on such a component.

The traditional manual calculation method used to assess a corroded structural member is by estimating the member net area after corrosion losses, verifying with applied load, and ensuring that member stress is not greater than allowable. Nowadays, the use of computer software such as SACS is popular, however the method only revealed the structural integrity of each member in the form of a unit check (UC); it did not indicate jacket structure global or system structural reliability. System structural reliability is vital for making appropriate decisions concerning an existing fixed platform structure's overall fitness for purpose. The method proposed in this textbook is based on a series reliability theory and parallel reliability theory respectively.

The formal definition of reliability is "the probability that a product will perform its intended mission for a certain period of time under a given conditions." (Summerville, Nicholas. 2004). Quality is defined as the totality of features and characteristics of a product or service that bear on its ability to satisfy a specific need or an implied need. Based on these definitions, there are four key factors of reliability that make it different from quality: probability, intended function, time, and environment. Therefore, a reliability method can be applied to structural engineering design and assessment guidelines to identify the members that are truly critical if additional members do improve the structure reliability. The reliability of a newly built structure is assumed to be 100 percent because the entire structural members and joints are expected to be brand-new and without corrosion dents. A typical aging and corroded structure with a corroded member would have a reliability value less than 100 percent. This proposal is based on these two assumptions.

The reliability assessment method for jacket structure presented in this section is established with sufficient knowledge of simple structural arrangement, known as parallel and series systems. The parallel ductile system is made up of members that share load equally, and the system fails when all the members fail, as illustrated in figure 9.9. *Ductile* means the members are perfectly plastic on failure, and the members cannot continue to support the load anymore.

In the case of a parallel brittle system, the failed member sheds load to the remaining members that have not yet failed in the system (see fig. 9.10). The corroded bracing structural member of jacket platform is a classic example of this scenario. The series or chain system requires only one member to fail for the whole system to fail; jacket leg is an example of this state. The series system performs the same way whether made of brittle or ductile members, as illustrated in figure 9.11.

For the fully ductile members system, the reliability increases with each member added. Conversely, the fully brittle member system reliability significantly reduces with the addition of new members. The next member has a high probability of failing when the load shed by the first failed member is transferred to the next member. The graph in figure 9.11 shows such systems in a more realistic correlation. The system also demonstrates that the highly indeterminate structures may not necessary be more reliable than a determinant structure.

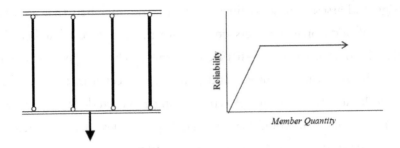

Figure 9.9 Parallel ductile simple systems

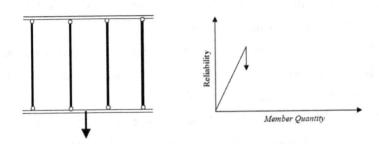

Figure 9.10 Parallel bristle simple systems

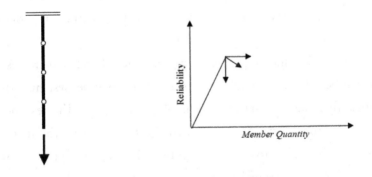

Figure 9.11 Series simple systems

For civil engineering, structures built from steel with certain member arrangements may demonstrate parallel ductile simple systems or series simple systems. The following characteristics are prevailing for the parallel ductile system.

- More members are better
- Correlation between members reduce this benefit
- Larger load uncertainty reduces this benefit

For series system, either ductile or brittle, the following characteristics are prevailing.

- More members are worse
- Correlation between members reduces this penalty
- Higher reliability of individual members reduces the penalty

The correlation between members in a system depends on whether the members are made from the same materials and whether whole members are subject to the same magnitude of load. For an independent member, the failure modes can be correlated as a result of sharing common members and more possible failure modes, but such a system does not necessarily indicate significantly lower reliability. Similarly, making a structure more statically indeterminate does not necessarily add much reliability. Therefore, whether a system is correlated, brittle, ductile, and parallel or in a series, the increase in individual components reliability is constantly elevating the reliability of the whole structures system. This is the most effective method available in engineering practice to improve structures' system reliability.

There are several methods applied to carry out the reliability and integrity of existing structures after corrosion losses. These methods include conventional and automated methods. The conventional method procedures estimate the member net area after corrosion wastage. Authenticate the residual member capacity by applying the theoretical acting load to calculate the member stress. The stress in a member must not be more than the allowable, as described mathematically in the following equations for a tubular section.

$$A_1 = \frac{\pi D^2}{4}\left(1 - \xi^2\right)$$ (9.1)

$$\xi = \frac{d}{D}$$ (9.2)

$$A_2 = \pi D \Delta t$$ (9.3)

$$A_o = \frac{\pi D^2}{4}\left(1 - \xi^2\right) - \pi D \Delta t$$ (9.4)

$$\frac{P_L}{A_o} \le \sigma$$ (9.5)

(A_1), (A_2), and (A_o) stand for member initial sectional area, the corrosion loss sectional area, and net sectional area, respectively; (d) is member internal diameter (mm), (D) is external member diameter (mm), (ξ) is the ratio between the internal and external diameter, (P_L) represents the applied load (kN), Δt is corrosion losses (mm), and σ is allowable steel stress (kN/mm²).

The equations provided above are valid only for a single structural member and cannot compute the structural reliability for a jacket platform with many structural members. Hence, the concept of structural reliability design has taken a more prominent position than the traditional deterministic design in the more advanced form. In the traditional design methods, parameters are used in deterministic values without uncertainties. In reliability methods, these values are not unique values but rather have probability distributions that reflect many uncertainties. For a classic marine structure, several uncertainties are obvious, such as fluctuations of loads and variability of material properties as applicable to a corroded member with thickness reduction.

For a corroded structural member, application of reliability theory with corrosion losses is based on the following equations.

$$R(t) = 1 - P_f(t) \tag{9.6}$$

$$R(t) = 1 - P_t(\Delta t) \tag{9.7}$$

$$R(t) = 1 - \frac{\Delta t}{T} \tag{9.8}$$

$$P_{f(\Delta t)} = \frac{\Delta t}{T} \tag{9.9}$$

Here, $R(t)$ is represent system reliability, $Pf(t)$ is the system failure, T denotes initial member thickness (mm), and (Δt) is the member corrosion losses (mm). Equation 9.7 can be rewritten in terms of member original thickness and time-variant member corrosion losses, as shown in equations 9.8.

9.8 Series System Reliability

The series reliability model states that the components of a system are connected in some form of fashion, one after the other (see fig. 9.12), and the system reliability of this system is the combination of the success probabilities of the member components. Accordingly, if one component preceding a component fails, then the entire system fails. A typical example of such a system in the field of engineering is a product pipeline or a fixed offshore jacket platform's support leg. When a segment of a pipeline is punctured due to corrosion and accidental damages, the entire pipeline system will be no longer perform the intended functions.

INPUT → P_A P_B P_C P_D → OUTPUT

Figure 9.12 Schematic of series reliability diagram

In this case, all the components are essential for the successful operation of the system. Therefore, the system reliability is the probability that all the components will function correctly. The system reliability calculation is represented in equation 9.10.

$$R(s)(t) = R(p_A).R(p_B).R(p_C).R(p_D) \qquad (9.10)$$

The system above is a series network where the system is nonredundant. Components P_A, P_B, P_C, and P_D must work for system success, and only one pipe needs to fail for the entire system to fail.

If R_A, R_B, R_C, and R_D represent the reliability or probability of the successful operation of components A, B, C, and D, and if Q_A, Q_B, Q_C, and Q_D represent the probability of failure of A, B, C, and D, then the success of the system (S) can be represented in terms of Boolean logic as follows.

$$S = A \cap B \cap C \cap D \qquad (9.11)$$

The reliability or probability of success of the systems is:

$$RS = R_A.R_B.R_C.R_D \qquad (9.12)$$

For "n" components, the series can be universally written as:

$$R_S = R_1.R_2.R_3.R_4 ---- R_n \qquad (9.13)$$

A characteristic of series systems is that the greater the number of the components, the lower the system reliability. The least reliable component in the system determines the overall reliability of the system.

9.8.1 Parallel System Reliability

The parallel reliability system is a system that is designed with redundant components. This is often adopted when reliability of some of the items in the system is insufficient, or when reliability of a system tends to be low as time progresses due to material degradation, as it is applied to civil engineering structures with several members. Parallel systems can be either active parallel or standby parallel. In an active parallel system, the whole components are active at all times. For a standby parallel system, some of the components will be standing by in a ready state, but they will not be engaged until the first parts fails.

Active parallel systems, where components A and B are active at all time, are illustrated in figure 9.13. A good example of this type of this system is the bracing members of a jacket structure with active redundant members.

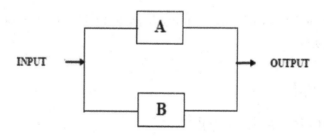

Figure 9.13 Active parallel system diagram

The system is believed to be operating at all times under one of the following conditions.

(1) A and B are both functional.

(2) Item A is functional, and B has failed.

(3) Item B is functional, and A has failed.

(4) If A and B fail, then the system is considered to be a failure.

The calculation for the reliability of the system, which is active parallel, is represented in equation 9.14.

$$R(s) = R(a) + R(b) - R(a).R(b) \qquad (9.14)$$

$R(s)$ is reliability of the system, and $R(a)$ and $R(b)$ are the reliabilities of each component in the system.

For the stand–by parallel, the system is fully redundant. Either A or B, or a combination of A and B, in the working condition will make the system successful. All components must fail for the system to fail. An example of this system may be a redundancy bracing member of a jacket structure. The failure of standby parallel can be represented in Boolean Logic as:

$$F = A \cap B \qquad (9.15)$$

The probability of the system failure is given by either equation 9.16 or equation 9.17

$$P_S = P_A.P_B \qquad (9.16)$$

$$P_S = 1 - \{(1 - R_A).(1 - R_B)\} \qquad (9.17)$$

P_A and P_B are the failure probability of items A and B, respectively, and R_A and R_B are the success probability of items A and B, respectively.

9.8.2 Jacket Structure System Reliability

The theory of parallel system reliability and series system reliability described in the earlier sections shall be followed step by step to carry out the calculation of jacket structure system reliability. The reliability flowchart presented in figure 9.14 will be the guideline. The jacket structure's member corrosion losses are presented in table 9.1, which shows as-built thickness in 1985 and the member thickness reduction in 2008 due to corrosion wastage.

The jacket structure's 3-D model in figure 9.15 shows the jacket legs and bracing arrangement with applicable elevations and coordinates. The bracing was grouped in accordance with the assumption of how the members are working together as parallel systems. But the jacket legs are considered to be working individuals in a series system.

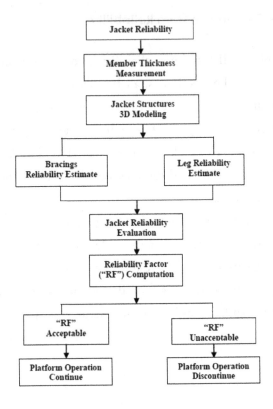

Figure 9.14 Jacket Reliability System Flowcharts

The values of member thickness corrosion losses obtained in the UT survey and presented in table 9.1 were used for the jacket structure reliability calculation in this section. The jacket bracings are divided into groups that are represented as A, B, C, D, E, and F. All the jacket bracings are assumed to be active, but some may be redundant or used below 100 percent capacity. Therefore, the jacket bracing system is believed to be operating at all times under one of the following modes.

1. Some bracing in the group failed, and others are functional.
2. All bracing groups (A, B, C, D, E, and F) are functional.
3. All bracing groups (A, B, C, D, E, and F) fail, and then the system is considered a failure.

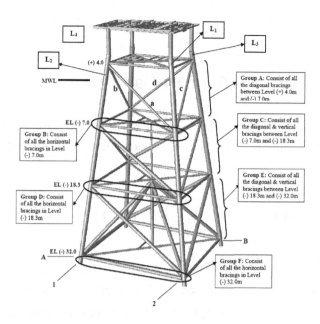

Figure 9.15 Jacket structures 3-D diagram

9.8.3 Jacket Structure Bracing Reliability

The jacket structures bracing members' position and arrangement vary and the bracings are grouped into A, B, C, D, E, and F based on individual bracing location (horizontally, diagonally, or vertically). It was expected that all the bracing members positioned in the same level and arranged in the same manner work together as a group in a parallel system. Therefore, every bracing in the same group works together and operates in an active parallel system principle. The entire bracing group is active at all times and is assumed to be operating under any of the following conditions.

(1) All the bracing groups are functional.
(2) Some members are operating, and a few members have failed.

If, and only if, all the bracing members at any of the group A, B, C, D, E, or F failed, then the group is considered to be completely failed (reliability

= 0), and the loads earlier carried by the failed group will be transferred to another group in the jacket structure system.

9.8.4 Jacket Structure Group Bracing Reliability

The jacket structures in figure 9.15 show group A bracings, which are arranged in parallel mode. Group A bracing reliability is expressed in equation 9.18.

$$R_A = 1 - \left[\left(P_a + P_b + P_c + P_{d.} - P_a.P_b.P_c.P_d \right) \right] \qquad (9.18)$$

R_A is the reliability of the bracing group A, and P_a, P_b, P_c and P_d are the failure probabilities of each bracing member or member thickness corrosion wastage. Accordingly, the reliability of other bracing groups B, C, D, E, and F will also be calculated.

The system failure is represented symbolically as F_f which is presented in a Boolean logic equation below.

$$F_f = A \cap B \cap C \cap D \cap E \cap F \qquad (9.19)$$

The probability of failure of the group bracing system is represented in equation 9.20.

$$P_{SG} = P_A.P_B.P_C.P_D.P_E.P_F \qquad (9.20)$$

$$P_{SG} = 1 - \left\{ (1 - R_A).(1 - R_B).(1 - R_C).(1 - R_D).(1 - R_E).(1 - R_F) \right\} \qquad (9.21)$$

$P_A = (1 - R_A)$, $P_B = (1 - R_B)$, $(1 - R_C)$, $(1 - R_D)$, $(1 - R_E)$, $P_F = (1 - R_F)$, which is group bracing failure probabilities.

The reliability of the group bracing system is given in equation 9.22.

$$R_{SG} = 1 - P_A.P_B.P_C.P_D.P_F \qquad (9.22)$$

9.8.5 Jacket Structure Leg Reliability

Jacket platform structures legs are working in series, and the leg system reliability is the product of all the leg reliability. Therefore, when one leg fails, the system fails because every leg is essential to the successful operation of the platform. Accordingly, the four-legged jacket system reliability calculation due to corrosion losses (R_{SL}) is presented in equation 9.23.

$$R_{SL} = R_1.R_2.R_3.R_4 \qquad (9.23)$$

$R_1.R_2.R_3.R_4$ - reliability of each jacket platform's four legs

9.8.6 Jacket Structure System Reliability

The jacket structure consists of four legs and several bracings. The bracings are grouped into A, B, C, D, E, and F in accordance to how they work collectively to resist the external loads. Based on the principle of parallel and series systems theories, the jacket structure's system reliability network reduction is formulated and presented in figure 9.16. R_A, R_B, R_C, R_D, R_E, and R_F represent reliability of group bracings that are arranged in a parallel manner, and L_1, L_2, L_3, and L_4 represent the jacket's four legs that are arranged in a series mode.

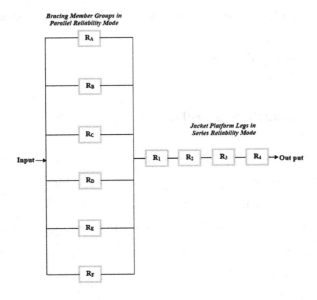

Figure 9.16 Jacket platform structure system reliability schematic

Jacket structures' system reliability can be defined as a product of group bracing reliability and jacket leg reliability, which is represented mathematically in the equation below.

$$R_{JS} = R_{SL}.R_{SG} \qquad \text{(9.24)}$$

R_{SL} represents jacket leg reliability, and R_{SG} denotes bracing group (A, B, C, D, E, F) reliability. A ratio known as reliability factor (RF) is established between reliability of a newly built jacket structure, which is assumed to be equal to 100 percent, and reliability of corroded jacket structure system, which is already less than 100 percent due to corrosion wastage. The proposed factor is represented mathematically in equation 9.25.

$$RF = \frac{1}{R_n} \qquad \text{(9.25)}$$

MATTHEW OMOTOSO, PH.D

R_n—corroded jacket structure system reliability

RF—reliability factor

9.8.7 Jacket Structure System Reliability Calculation Example

The jacket platform structural members' corrosion wastage data using ultrasonic devices method was presented in table 9.1. These data shall be used for the calculation of a jacket structure's system reliability as an example. The probability of failure and reliability probability were calculated for each bracing member using equation 9.9 and 9.8 respectively. The group bracing reliability was calculated for bracing group-A by applying equation 9.18. These calculations were presented in table 9.4. The total reliability of the six bracing groups (A, B, C, D, E, and F) was calculated using equation 9.22 and presented in table 9.5.

Jacket legs are operating in a series system, and the total sum of the jacket legs' reliability is calculated in accordance to equation 9.23 and presented in table 9.6. The total sum of jacket system reliability is obtainable in table 9.7 using equation 9.24.

The sum total of the investigated jacket structure system reliability (Rsj) is 0.85773, and the system reliability of a newly built jacket structure is 1.0. The reliability factor is calculated using equation 9.25 and is presented in table 9.7. The reliability factor is inversely proportional to the system reliability, as revealed in figure 9.19. When the system reliability decreases, the reliability factor of the system increases proportionally.

Table 9.3 Jacket member bracing and group bracing reliability

Group	ID	Corrosion Loss = tp (%)	Failure Probability (P = tp/100)	Reliability (1 - P)
A	1DA, (Pa)	0.262	0.00262	0.99738
	1D1, (Pb)	1.312	0.01312	0.98688
	1D2, (Pc)	4.462	0.04462	0.95538
	1DB, (Pd)	0.000	0.000	1.00000
	Reliability (R$_A$)	1- [(Pa + Pb + Pc + Pd) – Pa.Pb.Pc.Pd]		0.93964

Table 9.4 Summary of jacket bracing group reliability

Group	ID	Reliability (R)	Failure Probability P = (1 - R)
A	R$_A$	0.9396	0.06036
B	R$_B$	0.7717	0.22834
C	R$_C$	0.9475	0.05249
D	R$_D$	0.5879	0.41214
E	R$_E$	0.4102	0.58985
F	R$_F$	0.7850	0.21499
Reliability = R$_{SG}$	1 - P$_A$.P$_B$.P$_C$.P$_D$ P$_E$.P$_F$		0.99996

Table 9.5 Jacket legs reliability

Group	ID	Corrosion Loss = tp (%)	Failure Probability (P = tp/100)	Reliability (1 - P)
Support Legs	L01, (P$_{L1}$)	5.183	0.05183	0.94817
	L02, (P$_{L2}$)	4.450	0.0445	0.95550
	L03, (P$_{L3}$)	3.037	0.03037	0.96963
	L04, (P$_{L4}$)	2.356	0.02356	0.97644
	Reliability (R$_{SJ}$)	P$_{L1}$.P$_{L2}$. P$_{L3}$.P$_{L4}$		0.85777

Table 9.6 Jacket system reliability

Group	ID	Reliability (R)	Failure Probability P = (1 - R)
Support Legs	(R$_{SL}$)	0.85777	0.12150
Jacket Bracings	(R$_{SG}$)	0.99996	0.0001
Reliability (R$_{SJ}$)	R$_{SL}$.R$_{SG}$		0.85773

Table 9.7 Jacket system reliability and reliability factor

S/N	Period	1985	2008
1	Duration	0 year	23 years
2	Support Legs (R_{SL})	1.0	0.9995
3	Jacket Bracing (R_{SG})	1.0	0.8578
4	Reliability (R_{SJ})	(1.0 x 1.0) = 1.0	0.858
5	**Reliability Factor (*RF*)**	**(1.0/1.0) =1.0**	**1.0/0.8577) = 1.166**

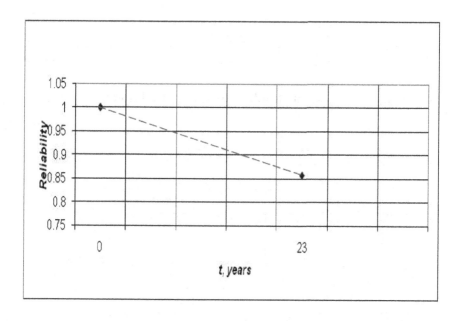

Figure 9.17 Jacket Structure System Reliability Schematic

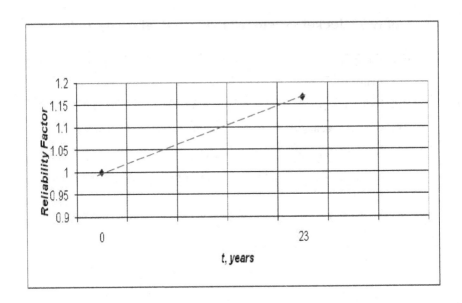

Figure 9.18 Jacket structure reliability factor versus age

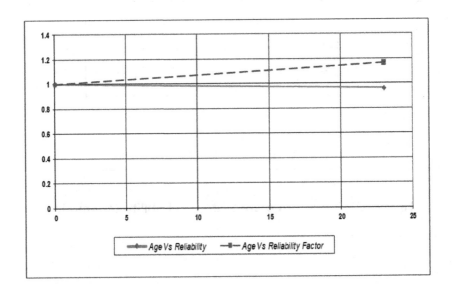

Figure 9.19 Jacket Structure Reliability Versus Reliability Factor

MATTHEW OMOTOSO, PH.D

9.9 Benefits of the Reliability Assessment Method

The jacket structure system reliability assessment method developed in this work is an appropriate technique that eliminates the rigorous exercises associated with complex computer analysis and manual calculations. Nevertheless, the available computer software and manual calculations methods are good for the new design and are not appropriate where structure system reliability is required.

The structure system reliability assessment method is a handy tool to monitor civil engineering structures safety against corrosion damage. The technique can be accomplished with a pocket calculator or Microsoft Excel. The benefit of the method includes provision of system reliability and individual member reliability that may not be readily accomplished by other available methods. This accomplishment is significant for the straightforward assessment of jacket platform structure, particularly when the platform life extension is anticipated. The study outcome shows that the cost of assessment using either manual or computer software is costlier than the system reliability method proposed in this book.

9.10 Conclusion and Recommendations

The reliability factor in this edition is an essential parameter to determine existing jacket structure safety, and the maximum value recommended for a fixed platform in active operation is 1.20. However, individual operators may fix the reliability factor values for their facilities based on the company's best engineering practices. The work revealed that more bracing members are better, but the correlation between the members reduces the system benefit. The work shows that fixed platforms with many legs are at a disadvantage because the series system exhibits that more members are worse, and a higher reliability of each leg reduces the penalty.

CHAPTER 10

CORROSION RISK-BASED ASSESSMENT

10.0 Introduction

Risk analysis is a systematic process of evaluating the potential risks that may be involved in a projected activity or undertaking. The analysis is a collection of several activities performed to provide support for decision making. A hazard is a condition that has potentially caused harm to people, property, and environment.

Risk assessment and management of safety, health, and environment protection is a significant part of design and construction in the oil and gas industry and beyond. This is a collection of several activities performed to provide support for decision making. The intentions of the risk screening process are to identify the high-risk areas in the systems and determine critical damage mechanisms that require detailed evaluation. The risk screening process is also engaged to classify various scenarios based on their frequency, failure probabilities, and consequences.

The source of a potential risk for typical engineering facilities is the deviation from its intended physical conditions. This includes insufficient mechanical strength due to corrosion or fatigue degradation and parameters out of acceptable range. A typical risk analysis flowchart is shown in figure

10.1. Risk means exposure to hazard, and risk level is determined by the severity of the consequences and the probability of an incident occurring. A general expression of risk R is described in equation 10.1.

$$R = \Sigma f(p, C) \tag{10.1}$$

Here, p and C denote frequency and consequence of incident, respectively.

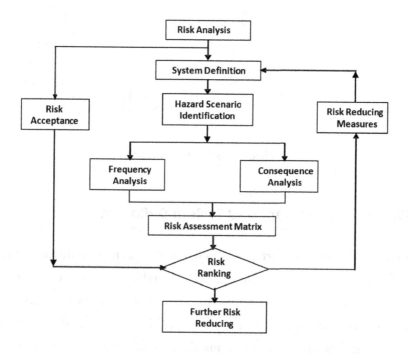

Figure 10.1 Risk analysis flowcharts

The flowchart summarily explains how risk-based assessment can be performed, and it offers recommendations for risk mitigation or even total elimination. To accurately carry out risk-based assessment, there should be a scenario development via event trees, which provide various event scenarios and probability factors. Such an example is illustrated in figure 10.2.

The measure of consequences is referred to as severity. Providentially, more severe events are not likely to occur more than minor events, and understanding the factors that differentiate a small event from a larger event help shape our safeguards. Probability represents the likelihood that the incident will occur within a given time frame. This can also be expressed as a number of incidents per period of time.

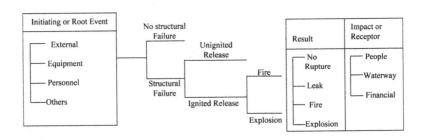

Figure 10.2 Event trees

10.1 Risk Assessment and System Definition

The risk analysis flowchart shown in figure 10.1 is appropriate for offshore facilities' structural assessment that are vulnerable to corrosion and fatigue damage. The risk ranking forms the basis for the selection of structure damage scenarios that may make the facilities be subjected to periodic inspection. The risk assessment matrix considered for the risk ranking scenarios is presented in table 10.1. The matrix is divided into three major regions, namely, unacceptable risk (A and B), acceptable risk (E), and the region between acceptable and unacceptable, which medium (C and D). The boxes in red, yellow, and white represent high risk, medium risk, and low risk, respectively. The criteria for offshore facilities structure hazard probability and damage consequences are presented in table 10.2 and table 10.3, respectively.

For thorough risk analysis, contribution from experts in the relevant field of engineering and safety, health, and environmental (HSE) specialist is apparently important. The dominant factors that are usually considered in consequence with analysis include the types of species that could be released into the environment and their associated hazards, the amount available for release, and the rate of release. Corrosion and materials engineering expertise is also required to estimate the amount and nature of damage that could happen to facilities.

In practice, the risk analysis process usually consists of group discussion and consultation with facility designer, operator, and maintenance team for proper identification of possible damaged scenarios. During such deliberation, scenario worksheets, consequences, and risk mitigations are adequately deliberated and resolved. At the implementation of these measures, the risk identified is expected to be reduced or totally eliminated. In the hierarchy of risk control, absolute elimination of exposure to risk is the best solution (see table 10.1).

Table 10.1 Risk Assessment Matrix

RISK ASSESSMENT MATRIX					
CONSEQUENCES	PROBABILITY				
	A	B	C	D	E
I					
II					
III					
IV					

Table 10.2 Industrial facilities damage consequence

Consequence Category	Health/ Safety	Public Disruption	Financial Impact	Environmental Impact
I	Fatalities or serious health impact on public	Evacuation of the whole personnel from the platform and continuing national or international attention	Corporate	Potential widespread, long-term, significant, adverse effects
II	Permanently disabling injury and serious lost time	Evacuation of the whole personnel from the facility and continuing regional attention	Business	Potential localized, medium-term, significant, adverse effects
III	Minor lost time injury with medical aid	Evacuation of some personnel and onetime regional attention	Field	Potential short-term, minor, adverse effects
IV	First aid	No evacuations, minor inconveniences to a few personnel	Others	Confined to lease or close proximity

Table 10.3 Hazard probability for industrial facilities

Probability Category	Definition	Interpretation
A	Possibility of repeated incidents	Industrial facilities with current conditions that indicate repeated future occurrences are possible
B	Possibility of isolated incidents	Industrial facilities with current conditions indicate several future occurrences are possible
C	Possibility of occurring sometime	Industrial facilities with current conditions indicate occasional future occurrences are possible
D	Not likely to occur	Industrial facilities with current conditions indicate future occurrences are not likely to occur
E	Practically impossible	Industrial facilities with current conditions indicate future occurrences are practically impossible

10.2 Offshore Jacket Platform Risk Analysis Framework

10.2.1 Hazard Scenarios Identification

Accidents often occur due to several possible minor failures that may create an unexpected weak mode in the structural systems and finally result in major structural failure. Potential corrosion and fatigue hazards damaged scenarios documented during recent offshore jacket platform inspections are listed below.

1. Member uniform corrosion (bracing and support leg)
2. Member localized corrosion (bracing and support leg)

3. Joint uniform corrosion
4. Joint little fatigue (nucleation period)
5. Joint medium fatigue (fatigue growth period)
6. No fatigue

The probable offshore jacket platform structure damage scenarios and initiators due to corrosion and fatigue damage are presented in table 10.4.

Table 10.4 Corrosion and fatigue hazard scenarios

Scenario	Description
Scenario 1	Joint and Member Uniform Corrosion + Little Fatigue
Scenario 2	Joint and Member Uniform Corrosion + No Fatigue
Scenario 3	Joint Localized Corrosion + Little Fatigue
Scenario 4	Member Localized Corrosion + Little Fatigue
Scenario 5	Member Localized Corrosion + Medium Fatigue
Scenario 6	Member Localized Corrosion + No Fatigue
Scenario 7	Joint Localized Corrosion + Medium Fatigue
Scenario 8	Joint Localized Corrosion + No Fatigue
Scenario 9	No Impact

(a) Little fatigue—nucleation period
(b) Medium fatigue—fatigue growth period
(c) Localized corrosion—pitting corrosion
(d) Uniform corrosion—general corrosion

10.2.2 Risk Probability Factor

The anticipated risk probability factor is based on the jacket platform structure's damage conditions, safeguards, and the past experience of the author, presented in table 10.5. The probability factors considered for performing the risk analysis process for the jacket platform structure damage include component vulnerability to corrosion, facility inspections, operator surveillance, and simultaneous damage of corrosion and fatigue hazard and redundancy components.

Nonetheless, these factors are dependent on the specific details of a project. The hazard safeguards vary in accordance to the nature of hazard and environment. The probability factors for a project could also include other considerations: decommission, impose load reduction, and periodic maintenance.

Table 10.5 Risk analysis probability factors

S/N	Risk Analysis Probability Factors
1	Corrosion failure
2	The operator noticed corroded component before failure
3	Operator surveillance
4	Redundancy structural members prevent failure
5	Combination of corrosion and fatigue hazard

10.2.3 Risk Analysis Tools

The risk analysis of a degraded jacket platform structure as a result of corrosion and fatigue hazard will perform based on the listed risk assessment tools.

1. Event Tree Scenarios
2. Scenarios Analysis
3. Scenarios Consequences
4. Qualitative Probability
5. Quantitative Probability

10.2.4 Event Tree Analysis (ETA)

This is a forward, bottom-up, logical modeling technique for both success and failure that explores responses through a single initiating event and lays a path for assessing probabilities of the outcomes and overall system analysis. This analysis technique is used to analyze the effects of functioning or failed systems given that an event has occurred. ETA is a powerful tool that will identify all consequences of a system that have a probability of occurring after an initiating event. With this forward logic process, use of ETA as a tool in risk assessment can help prevent negative outcomes from occurring by providing a risk scenario with the probability of occurrence. ETA uses a type of modeling technique called an event tree, which branches events from one single event using Boolean logic.

10.2.5 Scenarios Analysis (SA)

Scenario analysis is a process of analyzing possible future events by considering alternative possible outcomes. SA is a form of projection, but it does not attempt to show one exact picture of the future. Instead, it presents several alternative future developments and paths leading to the outcomes. The scenario analysis is not based on extrapolation of the past or the extension of past trends. Instead, it tries to consider possible developments and turning points that may be connected only to the past. Several scenarios are considered in scenario analysis to show possible future outcomes. Each scenario normally combines optimistic, pessimistic, more and less, or yes and no probable developments. Experience shows that few scenarios are most appropriate, and many scenarios seriously complicated the analysis.

10.2.6 Scenarios Consequences Analysis (SCA)

Scenarios consequences analysis can be defined as penalties that logically or naturally follow from an action or condition. This may be significance consequence or no consequence. Similar to scenario analysis, SCA does not depend on the extension of past trends, but it tries to consider possible developments and turning points. Numerous scenarios may be considered in SCA to show possible future outcomes. Individual scenarios normally combine optimistic, pessimistic, more and less, or yes and no probable developments.

10.2.7 Qualitative Probability

Qualitative probability risk analysis uses a relative or descriptive scale to measure the probability of occurrence. For example, a qualitative analysis would use a scale of low, medium, or high to indicate the likelihood of a risk event occurring. In a qualitative analysis, likelihood or probability is measured using a relative scale. Likelihood scale definition is presented in table 10.6.

Table 10.6 Likelihood scale

S/N	Likelihood	Description
1	Very Low	Highly unlikely to occur. May occur in exceptional situations.
2	Low	Most likely will not occurs. Infrequent occurrence in past projects.
3	Medium	Possible to occur.
4	High	Likely to occur. Has occurred in past projects.
5	Very High	Highly likely to occur. Has occurred in past projects, and conditions exist for it to occur on this project.

Note that these scales are dependent on the specific details of your project. For example, a low likelihood of occurrence for one project may mean a risk event is unlikely to occur within the next year. The impact scale for your project could also include other considerations such as scope, political factors, and employee impacts.

10.2.8 Quantitative Probability

Quantitative probability risk analysis is a quantitative analysis using a numerical scale. A quantitative analysis will determine the probability of each risk event occurring. For example, Risk 1 has an 80 percent chance of occurring; Risk 2 has a 27 percent chance of occurring, and so on. In summary, one might say that quantitative risk analysis breaks down risks from a high, medium, low ranking to actual numerical values and probabilities of occurrence to compute the overall effects.

10.2.9 Risk Analysis Summary

The risk analysis exercise carried out in this section, from figure 10.3 to figure 10.7, commenced with the preparation of an event tree in accordance with the established system definition. The scenario is arranged in such a way to match the event tree for the appropriate analysis. Each of the scenarios is simulated by specifying yes or no for event trees, scenario analyses, and consequences. Qualitative probability shows the level of probability of event occurrences (high, medium, or low), and quantitative probability provides quantitative values for every scenario (e.g., 90 percent or 70 percent is applied to highly likely scenarios, and 30 percent or 10 percent is use for low likely scenarios).

The risk analysis of the condition consequence gives associated risk levels for each scenario in the consequence hierarchy (I, II, III, and IV). The risk level for each scenario was indicated in a risk assessment matrix (table 10.7). Scenario 7 falls within the high-risk zone in the RAM, and therefore it is considered as a high-risk state among the other scenarios. Therefore, several hazard preventive measures are suggested to reduce the risk level of scenario 7, as well as the other scenarios.

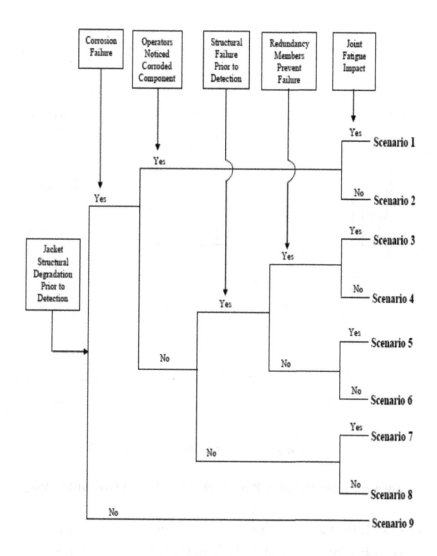

Figure 10.3 Event tree

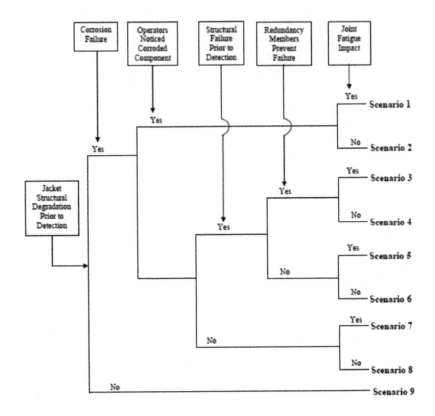

Figure 10.4 Scenarios outcome

Consequences (Health, Public, Financial, and Environmental)

Scenario 1: Joint and Member Uniform Corrosion + Little Fatigue

Scenario 2: Joint and Member Uniform Corrosion + No Fatigue

Scenario 3: Joint Localized Corrosion + Little Fatigue

Scenario 4: Member Localized Corrosion + Little Fatigue

Scenario 5: Member Localized Corrosion + Medium Fatigue

Scenario 6: Member Localized Corrosion + No Fatigue

Scenario 7: Joint Localized Corrosion + Medium Fatigue

Scenario 8: Joint Localized Corrosion + No Fatigue

Scenario 9: No Impact

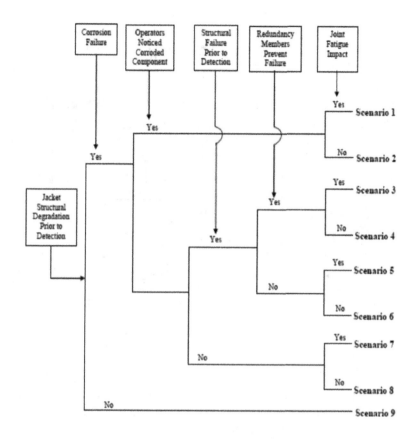

Figure 10.5 Scenarios consequences

Consequences (Health, Public, Financial, and Environmental)

Scenario 1: Joint and Member Uniform Corrosion + Little Fatigue (III)

Scenario 2: Joint and Member Uniform Corrosion + No Fatigue (IV)

Scenario 3: Joint Localized Corrosion + Little Fatigue (II)

Scenario 4: Member Localized Corrosion + Little Fatigue (III)

Scenario 5: Member Localized Corrosion + Medium Fatigue (II)

Scenario 6: Member Localized Corrosion + No Fatigue (III)

Scenario 7: Joint Localized Corrosion + Medium Fatigue (II)

Scenario 8: Joint Localized Corrosion + No Fatigue (III)

Scenario 9: No Impact (Not Applicable)

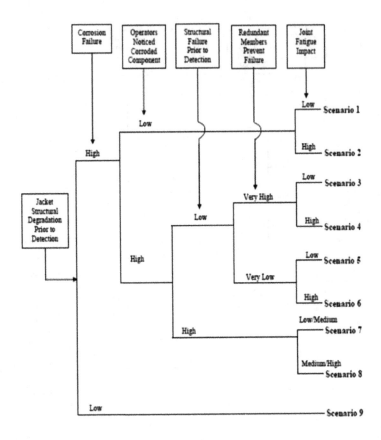

Figure 10.6 Qualitative probability

Scenario 1: Low   Scenario 2: Medium   Scenario 3: Low

Scenario 4: Medium   Scenario 5: Very Low   Scenario 6: Low/Very Low

Scenario 7: High   Scenario 8: High   Scenario 9: Medium

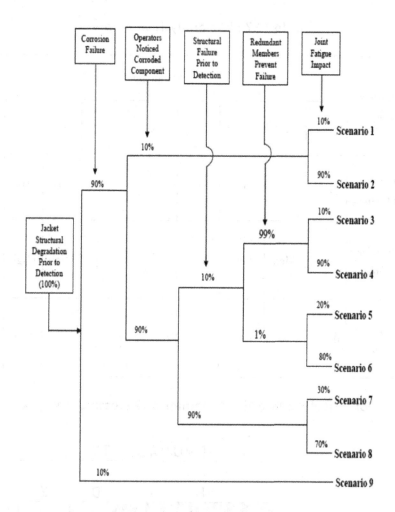

Figure 10.7 Quantitative probability

Scenario 1: 0.00900   Scenario 2: 0.08100   Scenario 3: 0.00800
Scenario 4: 0.07200   Scenario 5: 0.00016   Scenario 6: 0.00064
Scenario 7: 0.21900   Scenario 8: 0.51000   Scenario 9: 0.10000

Table 10.7 Risks analysis summary

Probability Factors	Scenarios								
	1	2	3	4	5	6	7	8	9
Scenarios Consequences	III	IV	II	III	II	III	II	III	N/A
Qualitative Probability	Low	Med	Low	Med	Very Low	Low/ Very Low	High	High	Med
Quantitative Probability	0.009	0.081	0.008	0.072	0.0002	0.0006	0.219	0.51	0.10

Table 10.8 Plotting Risk Scenarios on Assessment Matrix

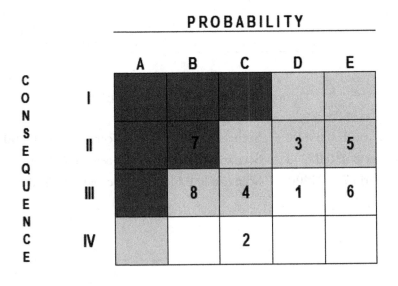

MATTHEW OMOTOSO, PH.D

10.3 Hazard Preventive Measures

As documented in table 10.2, industrial facilities failure include offshore jacket platform has a negative impact on HSE, public disruption, and the finances of the stakeholders. Scenario 7 (joint localized corrosion and medium fatigue) falls within the high-risk zone in risk assessment matrix (RAM) in table 10.8, requires effective hazard mitigation measures to urgently reduce the risk level. The jacket offshore platforms in question are by standard design with a cathodic protection system and coating as a safety mitigation measure against corrosion and fatigue hazards.

Based on the outcome of a risk assessment workshop conducted, the following additional mitigation measures were recommended by the participants to reduce scenario 7's risk levels.

1. Cathodic protection system renewal
2. High corrosion-resistant structural steel
3. Provision of corrosion allowance
4. High-yield-strength structural steel
5. Good quality welds
6. Adherence to platform design life (25–30 years)
7. Reduction of load on the platform

After the assumption that the suggested mitigation measures have been implemented, the workshop participants carried out the risk analysis review on scenario 7, and it was established that scenario 7 risk level has been significantly reduced, as detailed in table 10.9.

Table 10.9 Scenario 7 risks ranking

RISK RANKING							
BEFORE RECOMMENDATIONS				AFTER RECOMMENDATIONS			
S_b	P_b	E_b	F_b	S_a	P_a	E_a	F_a
B-II		D-II	C-II	E-II		E-III	E-III

S—safety; P—public disruption; E—environmental; F—financial;
b—before recommendations; a—after recommendations

10.4 Conclusion

The selection of appropriate material during the engineering design phase can hardly be overemphasized for corrosion and fatigue hazard mitigation. This should be done in an accord between technical competence and economic factors of the substance with due consideration to the mechanical properties, corrosion resistance, service temperature, and chemical resistance, among others. These factors are well represented and guided with industrial standards. For repair purposes, there may be fewer opportunities for the new material selection. Therefore, the major factors will be centered on the fabrication simplicity and remaining life span of the facilities to avoid overdesigning when considering corrosion allowance.

The outcome of the risk analysis demonstrates that offshore jacket structures required survey rules in order to control corrosion and fatigue damages. Offshore jacket platform structures required frequent inspection, and substructure parts with limited access for inspection gave been noted for fatigue damage. These parts are recommended for high fatigue safety factors during the design stage.

MATTHEW OMOTOSO, PH.D

Emphasis should be laid on cathodic protection system revalidation as the most cost-effective solutions for the protection of marine structure against corrosion. Experience recognized the high cost, logistics, and environmental concerns involved in replacing depleted anodes. Therefore, a non-weld retrofit option that cuts down on cost and reduces installation time is recommended.

Structural joint construction improvement using a joint can to replace the usual simple joint without overlapping braces is highly recommended for marine structures to prevent joint failures. The outcome of risk analysis has given practical application to marine structure assessment against failures due to corrosion and fatigue hazards.

CHAPTER 11

CORROSION SAFETY RISKS AND ECONOMICS

11.0 Introduction

NACE study estimates global cost of corrosion at $2.5 trillion annually This corrosion cost represents a significant part of individual country gross national product (GNP). The real issue of this annual corrosion cost is that some amount could be saved by the application of existing technology to prevent and control corrosion.

Corrosion protection not only maintains the value of facilities but also ensures safe operational conditions and reliability of the structures. It averts damage that can endanger people and the environment. Hence, corrosion protection of engineering installations forms the most important means of protecting the environment. Corrosion-related failures of facilities are key sources of risks. Pitting or environmentally assisted cracking corrosion can be a life-limiting cause of installations' deterioration, leading to loss of containment of hydrocarbon fluids and general structural failure.

11.1 Corrosion Safety Risks

Two characteristics apply to petroleum products: the inherent nature of being flammable and volatile. Flammable means that they can burn under certain conditions. Volatile means that they can vaporize when unconfined, and the vapor can form an explosive mixture when combined in suitable proportions with oxygen. Pipelines transport flammable products, and their operators do everything possible to protect the safety of the facilities by cooperating with technical societies and governmental authorities.

Several petroleum product pipeline accidents have been reported in different parts of the world. The immediate causes of these catastrophes are due to pipeline ruptures that could be attributed to many factors, such as corrosion damages, terrorist acts, and pipeline vandals and trespassers. The likely cause of most pipelines accidents is attributed to pipe wall thickness reduction due to severe internal corrosion. The excess corrosion might have occurred due to corrosion protection devices to prevent, detect, and control internal corrosion within the pipeline systems. The failure of a pipeline due to internal corrosion can be devastating. Control of internal corrosion in a steel pipe could be of significant benefit to pipeline operators. Studies show that several pipelines accidents could be attributed to one or a combination of the following reasons.

- Overpressure of pipeline, and external damages to the pipeline through excavation and external corrosion of the pipeline
- Pipe rupture caused by the combination of microbes and other contaminants such as moisture, chlorides, and hydrogen sulfides within the pipeline
- Lack of adequate cleaning pigging operation to thoroughly remove liquids and solids from the pipeline, leading to corrosion damage
- Little or no available internal corrosion control program for the internal corrosion that may occur in the pipelines

The operator is required to establish a continuing education program for the public to recognize a petroleum product pipeline emergency and report it to firefighters and the appropriate officials. In order to lessen the damage, the same campaign program should be extended to persons engaged in excavation-related activities for quick response to emergency conditions. However, emergency response procedure and capability are highly necessary for immediate reaction with equipment and personnel in case of any fire incident. For the areas prone to terrorists and pipeline vandals, the operators are strongly advised to put in place an effective pipeline surveillance twenty-four seven, watching over the pipelines against possible assault.

Hazards identification and risks ranking are fundamental for the management process of corrosion control and safety. A hazard has the potential to cause damage to installations, and risk is the combination of severity and probability of failure occurrence. For example, corrosion-related failures could lead to pipeline leakage, hydrocarbon releases, and loss of life. Facilities' risk assessment is a cautious assessment of potential hazards that may affect the smooth operation of a production. These may be risks associated with the safety integrity of structural members, vessel leakages, and poor corrosion mitigation procedures. Management of corrosion is therefore a major driver for personnel and environmental safety, as well as economic issues within the industry. Therefore, it is highly necessary to provide the appropriate and clear inspection methods that are considered to be a good industry practice.

Structure members and joints due to corrosion damage are frequently monitored by recognized inspection procedures, such as ultrasonic testing. In the case of pipelines and processing systems, the rate of corrosion is controlled by injection of inhibitor chemicals. The external surface of installations is naturally protected by corrosion-control coatings and in conjunction with cathodic protection systems for a very aggressive environment. Selection of appropriate materials of construction for corrosion in aggressive condition ensures inherent safety, reduces cost of maintenance, and lengthens installation life span. The availability of numerous national and international standards and codes has provided a general and progressive outline for industries' activities with specific duties falling on designers, constructors, and operators. That is, the facility operators must have effective plans and control, monitor, and preventive measures to secure the safety of their facilities against corrosion damage.

Another major concern is corrosion damage and hydrocarbon released in separation equipment and processing plants that are densely packed, as well as being exposed to both internal and external corrosive environments. The hazards associated with corrosion must be recognized, and considerable resources should be directed toward managing the risks. Most practices and procedures engaged for corrosion control, as specified in this book, are proven technology and are generally accepted worldwide. The existing corrosion control methods include materials selection, chemical treatments, coatings, cathodic protection, and design and environmental control. These options are used either individually or in combination; the choice depends on the specific application and the corrosivity of local environments. Engineering success requires selection of the most viable options using both technical and economical methods. By means of corrosion inspection and monitoring, combined with suitable maintenance strategies and procedures, the lasting integrity and durability of engineering installation can be achieved. Generally, there is a need to improve and advance the feedback route from operational knowledge and the lesson learned to future designs. This could be achieved by the provision of a direct input into engineering projects from operational personnel and experienced site engineers, who should conduct design audits and fabrication procedures.

The policy development for corrosion control is complex, especially the overall management of corrosion risks, the effective deployment of human resources, and the development of appropriate systems to meet the challenging conditions. Achieving a specified purpose in corrosion control requires codes, standards, and specifications with appropriate management procedures. The responsibility of day-to-day corrosion control and management could be shouldered by different groups and individuals. The strategy and interface between these individuals and groups are vital for the overall desired result of corrosion control strategies.

Corrosion control and mitigation could be achieved using monitored data that can be converted into management information and followed swiftly with good management decisions. The accomplishment of corrosion control is known to be judged against predetermined performance requirements and established standards. Reputable corrosion monitoring systems include active corrosion monitoring, uses of regular checks and inspections, and continuous evaluations to ensure that agreed criteria are met. Also, reactive corrosion monitoring methods engaged the recording of data after failure examinations and revamped works with other data that contribute to corrosion control under performance. A well-thought corrosion management flowchart showing corrosion management stages has been illustrated in figure 11.1.

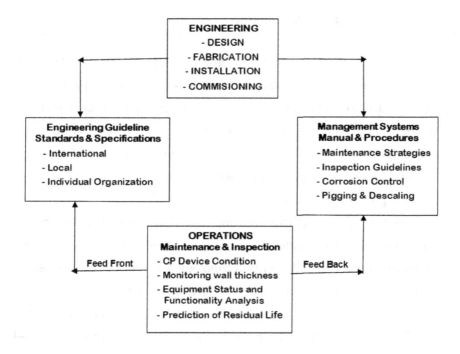

Figure 11.1 Corrosion management flowchart

The earlier mentioned monitoring systems require supporting procedures that not only investigate causes of substandard performance but also recommend improvements in procedures. Substandard performance must be investigated and reported for improvements by using standard reporting forms. Virtually all the industries, particularly offshore oil and gas production, are widely engaged with well-established corrosion mitigation procedures. Despite that fact, unacceptable issues such as leaks and structural failure still occur at their facilities. The opinion of the stakeholders is that the causes of corrosion-related failures are human error and poor management control. Another reason for failure includes lack of regular inspection, negligence of early warnings, and lack of technical information. For a lasting corrosion control, an overall system required managing not only technical corrosion issues but also human response and actions.

11.2 Economics of Corrosion

Corrosion is essentially an economic problem, and it is very important that engineers and facilities managers be aware of the economic impact on their business decisions in order to meet corporate goals. The cost of metallic corrosion will be considered from two viewpoints, the cost to the nation's economy and the cost of selected corrosion control measures. The average cost of corrosion to a typical nation is equivalent to 3–4 percent of its gross national product (Von et al. 1997).

The real calamity of this avoidable corrosion cost is that some of the corrosion-damaged facilities could be saved by the application of existing corrosion prevention and control technology. For years, many completed and operated facilities attest to the successful use of the existing technology for corrosion control that lessens the annual losses to corrosion. This existing technology includes coating, linings, cathodic protection, and inhibitors, among others.

However, prior to the selection of a corrosion control method for a particular project, the corrosion engineer should bear in mind the economic factors of the method. The corrosion method economics evaluation includes the technical validity and economic justification of the option. Nevertheless, a particular corrosion problem may have more than one solution that adequately solves the technical problem but with different cost implications. After assessment of technically viable alternatives and the associated cost, the performance characteristics should be considered as well for any selection that reduces the magnitude and time value of money.

Engineering economy is known to be the discipline that assists in measurement of the economic impact on engineering decision in order to meet the corporate goals. Engineering economy provides the economic techniques in a manner that can be understood and used by engineers as a decision-making tool. The subject also facilitates the communication of decisions between the engineer and corporate management to justify investment in a particular corrosion prevention method that has long-term benefits.

For instance, in the case of galvanic cathodic protection system, the choice of sacrificial anode during the design calculation phase may be arbitrary. Several anodes should be considered during design calculations for economic analysis reasons, as recommended in DNV-RP-B401. The cost and financial side of cathodic protection depend on a multiplicity of several parameters, making common views on costs not really feasible. The amount of current protection required and the exact electrolyte resistance in an installation environment with the selected anodes may vary significantly, leading to the cathodic protection system's cost. As an approximate estimate, the installation cost of cathodic protection of uncoated metal structures is about fifteen times greater than the coated surfaces. Therefore, it is highly advised that metal structures and pipeline should be coated prior to installation of cathodic protection systems.

The use of cathodic protection is by and large accepted today more than ever before due to the techniques and advantages for controlling some forms of corrosion that are cheaper than the other methods. Nonuniform corrosion damage like pitting, which frequently occurs on marine steel structures, leads to damages in tanks, pipelines, and ships and could be controlled and protected by a cathodic protection system. As a result of cathodic protection application, many product tanks were able to be constructed with low-alloy and high-strength steel instead of corrosion-resistant, high-alloy steel (which is expensive) with poor mechanical properties.

In the determination of the cost of cathodic protection for structures and pipelines, practical experience shows that the costs of the system will increase with the amount of current protection requirement, which depends on the amount of current leakage and the protected surface area. For impressed current system, the electrical soil resistivity at the anode location determines current level output from anodes and the subsequent rectifier output voltage required to supply the protection current. The distance between the nearest power supply and the impressed current anodes is also a factor when calculating the cost of an impressed current cathodic protection system. The higher the cost of linking the current of the impressed current system, the more economical the galvanic anodes system becomes. The selection of a corrosion protection method is decided not only by economic reason but also by technical efficiency considerations.

Marine structure splash zone and subsea pipelines' wall thickness are often increased to meet the planned service life, leading to a considerable increase in materials used. The increase of the construction materials consequently added to the structure weight, which increases the transportation cost, and the construction process takes longer. Therefore, provision of additional wall thickness regarding economical and technical advantages should be carefully considered with any other available corrosion protection methods to meet the planned service life of the structure.

Corrosion management includes many activities during the service life span of facilities that are performed to mitigate corrosion, which includes repairing corrosion-induced damage and replacing parts that become unusable due to corrosion damage. The total cost of corrosion is divided into two main categories, direct and indirect costs. The direct costs are the costs directly incurred by the facilities' owner, and indirect costs are the one incurred by others that are not directly provided by the facilities' owner. The recommendation is that the actual cost of corrosion should also include indirect costs in the analysis of corrosion's yearly cost alternatives, such as the cost of service interruption and environmental damage. A corrosion management plan that is considered a direct-cost cash flow is frequently applied on projects. For example, the breakdown of true cost of a coating system over its entire service life consists of original costs, touchup costs, maintenance costs, repainting costs, inflation, and opportunity costs. Using the standard formula for the annual maintenance cost, the current discount annual value {AM} can be expressed as follows.

$$PDV\{AM\} = AM\left(\frac{1-(1-i)^{-N}}{i}\right) \tag{11.1}$$

Here, AM is the cost of annual maintenance, N is the length of service life in years, and i is the interest rate.

The equation of net present value (NPV) for the cost of corrosion control options are taken in parametric form as follows.

$$NPV = \frac{C_c(1+i)^n}{(1+r)^n} \tag{11.2}$$

C_c is the current cost, i is the inflation rate, n is the lifetime of the project (years), r is the interest rate, and NPV is the net present value.

The above equations demonstrate that the choice of the corrosion control method should depend not only on initial cost but also on the yearly maintenance cost. It is more cost-effective to specify a material that provides an extended life in an area that is very difficult to perform repair works or in components that would cause major shutdowns in case of failure, as it is applicable in oil and gas subsea production equipment.

There are several options that may be considered during equipment selection, such as a low initial cost system largely based on carbon steel and cast iron that requires considerable maintenance over the life span of a plant. Nevertheless, this option may be a reasonable choice for an area where labor costs are low, material is readily available, and installations are easily accessible. There may be another choice of equipment, such as a system based largely on alloy materials, properly designed and fabricated with minimum maintenance and reliable function. Beyond a doubt, this alternative will be appropriate to the trend of rising labor costs in most industries and the need for high reliability in a capital-intensive plant.

In practice, many systems are a combination of these extreme options leading to high initial costs of one item and the high maintenance costs of another. Corrosion-resistant alloys such as stainless steels, titanium, and nickel-based alloys are used where corrosive conditions prohibit the use of carbon steels. Also, the protective coatings that could have been provided for carbon steel are insufficient protection and not economically feasible. Examples of this condition are engagement of stainless steels, nickel-based alloys, and titanium alloys used for the construction of subsea valves, tubing hangers, and sea fasteners. Nickel-based alloys and titanium alloys are particularly used in severe and aggressive environments such as refinery facilities and chemical-process industries, where high temperatures and corrosive conditions exist. Titanium alloys are used predominantly in the aerospace and military industry as a result of the alloy's advantages when it comes to its high strength-to-weight ratio and its sufficient resistance to high temperatures.

11.3 Depreciation

Depreciation has been defined as the lessening value of an assent with the passage of time. Most of the physical asset depreciates with time, except land for construction purposes. There are two common types of depreciation categories, physical depreciation and functional depreciation.

Physical depreciation refers to asset depreciation due to incidence of corrosion, rotting of wood, bacterial action, chemical decomposition, and wear and tear that reduces the asset capability to render its intended functions. Functional depreciation is not due to asset inability to serve its intended purpose, but rather due to availability of some other asset that can more economically and efficiently perform the desired function. Hence, inadequacy to meet the demands placed on an asset can lead to functional depreciation. Also, high-tech advances produce developments that often lead to obsolescence of existing assets. There are several methods in which asset depreciation can be calculated.

- Straight line
- Declining balance
- Declining balance switching to straight line
- Sum-of-the-year's digits
- Accelerated cost-recovery system.

The manner in which depreciation is accounted for is largely based on a government's tax regime. Taxes change from time to time, and for that reason, procedures that are attractive under a given condition may be a forbidden under other conditions. Thus, for the purpose of calculating asset depreciation as a result of corrosion damage, straight line and sum-of-the-year's digits methods shall be discussed further in this chapter.

In straight line depreciation method, depreciation is charged uniformly over the life of an asset that is under consideration. First subtract the residual value of the asset from its original cost to obtain the depreciable amount. The depreciable amount is then divided by the useful life of the asset in the number of accounting periods to obtain depreciation expense per accounting period. Owing to the simplicity of the straight line method of depreciation, it is the most commonly used. The formula to calculate the straight line depreciation of an asset for the complete period under consideration is expressed as follows.

$$D = \left(\frac{P - S}{n} \right) \tag{11.3}$$

D is the annual depreciation, P is the asset original cost, S is the salvage value, and n is the expected life span.

For example, a company purchases equipment costing 5,000,000 NGN. The equipment is expected to have a value of 800,000 NGN after ten years when it will be sold as obsolete equipment. Calculate the depreciation expense on the equipment when it will be seven years old.

First, find the depreciable amount, which is 4,200,000 NGN (5,000,000 NGN cost minus 800,000 NGN).

Then, divide the depreciable amount by 10 years, which is the useful life of the equipment, as shown in table 11.1. This will give 420,000 NGN for yearly depreciation. The depreciable amount in the seventh year will be 7 multiplied by 420,000, which equals 2,940,000. Therefore, if the company decided to sell the equipment at year seven, it will go for 5,000,000 minus 2,940,000 NGN, equal to 2,060,000 NGN.

Table 11.1 Asset straight line depreciation schedule

Period	Duration	Yearly Depreciation	Cumulative Depreciation
Year	1	420,000	420,000
Year	2	420,000	840,000
Year	3	420,000	1,260,000
Year	4	420,000	1,680,000
Year	5	420,000	2,100,000
Year	6	420,000	2,520,000
Year	7	420,000	2,940,000
Year	8	420,000	3,360,000
Year	9	420,000	3,780,000
Year	10	420,000	4,200,000
Equipment Value after 10 Years			800,000
Equipment Original Cost (4,200,000 + 800,000)			5,000,000

The sum-of-the-year's digits method considers that the value of an asset decreases at a decreasing rate. It is considered to be appropriate to use an accelerated depreciation method when an asset loses most of its value during the beginning of its useful life as it is applicable to automobiles.

This method takes the asset's expected life and adds together the digits for each year. If the asset was expected to last for six years, the sum of the year's digits would be obtained by adding 6 + 5 + 4 + 3 + 2

+ 1 to get a total of 21. After that, each digit is divided by this sum to determine the percentage by which the asset ought to be depreciated each year, commencing with the highest number in year one. For the six-year example given above, this method would yield the depreciation schedule presented in table 11.2.

Table 11.2 Asset sum-of-the-year's digits depreciation schedule

Period	Duration	Ratio	Percentage	Cumulative Percentage
Year	1	6/21	28.57	28.57
Year	2	5/21	23.81	52.38
Year	3	4/21	19.05	71.43
Year	4	3/21	14.29	85.71
Year	5	2/21	9.52	95.24
Year	6	1/21	4.76	100.00

The general equations that provide a simple solution for a large percentage of engineering economy problem have been provided in equation 11.4. Equation 11.5 is for the straight-line depreciation, which takes into account the influence of taxes, depreciation, operation expenses, and salvage value in the calculation of present worth and annual cost. When using this equation, each problem can be solved by entering data into the equation with the support of compound interest from tables 11.3. The tax rate, t, is represented as a decimal, and P is the cost of initial investment at time zero.

$$(PW) = -P + \frac{t(P-S)}{n}(P/A, i\%, n) \tag{11.4}$$

$$(PW) = -P + \frac{t(P-S)}{n}(P/A, i\%, n) - (1-t)(X)(P/A, i\%, n) + S(P/F, i\%, n) \tag{11.5}$$

The annual amount of tax credit permitted by this method of depreciation, in this straight-line depreciation, is expressed by [t(P- S)/n]. These equal annual amounts are translated back to zero time by converting them to present worth using the (P/A, i%, n) value from table 11.3. A work example follows.

Question 1: A new process vessel is required during the upgrade of an existing petroleum processing facility in offshore Niger Delta. Because of corrosion, the expected life of a carbon steel process vessel is five years, and the installed cost is 1,500,000 NGN. There was another proposal during the HAZOP meeting to substitute the steel process vessel with AISI 316 stainless steel, which has an installed cost of 4,500,000 NGN and an estimated life of fourteen years, written off in ten years. Which of these two options is more economical choice based on annual cost? The minimum acceptance rate of return is 10 percent, the tax rate is 35 percent, and the depreciation method is a straight line.

Answer 1: Because the lives of the alternatives are not equal, the economic choice cannot be based on the discounted cash flow over a single life of each alternative. Therefore, comparison should be made based on the basis of equivalent uniform annual costs. Application of equation 11.4 will be appropriate because the third term (involving maintenance expense) and the fourth term (involving salvage value) in equation 11.5 are both assumed to be zero.

$$\left(PW\right)_{steel} = -1,500,000 + \frac{0.35\left(1,500,000 - 0\right)}{5}\left(3.7908\right) = 1,101,966 NGN$$

$$\left(PW\right)_{Stainless} = -4,500,000 + \frac{0.35\left(4,500,000 - 0\right)}{10}\left(6.1446\right) = 3,532,226 NGN$$

To evaluate the two alternatives (steel and AISI 316), the discounted cash flows (PWs) of each is converted to annual cost.

$$(PW)_{steel} = 1,101,966(0.2638) = 290,699 NGN$$
$$(PW)_{316} = 3,532,226(0.1628) = 575,046 NGN$$

Therefore, the carbon steel process vessel with the lower annual cost is more economical alternative under these conditions.

Question 2: Assuming the carbon steel vessel was required with the sum of 500,000 NGN for maintenance in the form of painting and employing inhibitor and cathodic protection, would carbon steel still be the preferred alternative after one year?

Answer 2: The costs of maintenance are expenses of the item at the end of each year. This exercise can be solved by making use of the third term in equation 11.5. The annual cost, plus maintenance, is calculated by subtracting the after-tax maintenance cost from the answer in example 1.

$$A_{steel} = -290,699 - (1-0.35)(500,000) = -615,699 NGN$$

After comparing this amount (-615,699 NGN) with the annual cost of 575,046 NGN for A316 in example 2, it is very obvious that if 500,000 NGN is needed yearly for the maintenance of carbon steel vessel in order to keep it in good state, then A316 unit would be more economical.

Question 3: Considering the same situation explained above, how much product loss (X) is expected to tolerate after three of the five years of expectant life from roll leaks and other types of minor failures before choosing of A316 is reasonable?

Answer 3:

$$A_{316} = A_{steel} + A_{productloss}$$

$$-575,046=[290,699+(1-0.35)(X)(0.7513)](0.2638)$$

$$-575,046=76,686.396+0.1288(X)$$

$$X = -5,060,034 NGN$$

The term (1 - 0.35) is acknowledged from term three of equation 11.5. The quantity 0.7513 is the single-payment PPW factor (P/F, 10%, 3 years) in table 11.3, and it translates the product loss, X, to its present worth at time zero. The quantity (0.2638) is the uniform series capital recovery factor (A/P, 10%, 5 years). The present worth of the product loss is translated to the annual cost and added to the annual cost of steel for comparison with the annual cost of an A316 vessel. When the production losses exceed 5,06,034 NGN, the use of an A316 vessel is justified.

Question 4: There was a proposal to cathodically protect a crude oil pipeline to be built in a swamp area of Niger Delta, Nigeria, with the technical possibility of using either sacrificial anode or impressed current based on the following assumptions: $i = 10\%$, $t = 30\%$, and straight depreciation is recommended. There are two design options, described below, to determine the less expensive alternative.

Option A: Provide 75 sacrificial anodes and installation at a cost of 19.5 million NGN. Useful expected life is 20 years. Inspection of device is limited, hence no maintenance cost.

Option B: Require provision and installation of 34 groundbed impressed anodes and rectifier with a capacity of 126.97 volts and 16.16 watts at a cost of 15.5 million NGN. Expected life is 25 years with an annual maintenance cost of 850,000 NGN.

For Option A:

$$(PW)_A = -19,500,000 + \frac{0.30(19,500,000-0)}{20}(8.5136) = -17,009,772\,NGN$$

$$A_A = -17,009,772(0.1175) = 1,998,648.21\,NGN$$

For Option B

$$(PW)_B = -15,500,000 + \frac{0.30(15,500,000-0)}{25}(9.0771) - (1-0.3)(850,000)(9.0771)$$

$$= -15,500,000 + 1,688,340.6 - 5,400,874.50 = 19,212,533.9\,NGN$$

$$A_B = -19,212,533.9(0.1102) = 2,117,221\,NGN$$

Considering the above two scenarios for the pipeline cathodic protection, Option A is calculated to be the less expensive option on the basis of lowest annual cost.

Within the industrial sector, there is a range of current practices of dealing with corrosion protection, from the old technology to the current state-of-the-art. But, the goal of corrosion management is to achieve the desired level of service at the least cost. Capital budgeting techniques represent a controlling practice for evaluation corrosion protection engineering in terms of financial value to an organization. Finding the corrosion management program that has the greatest net benefits requires a careful analysis of all the direct and indirect costs involved. This analysis requires specific corrosion-related cost information to identify the design-maintenance option that had the lowest annual cost. Unfortunately, all too often, selection of materials for corrosion applications is still solely based on a comparison of initial installed costs of alternative materials. The time value of money concept, considerations for ease of repair, costs associated with planned and unexpected shutdowns, and the effect of component failure on overall plant operations are thereby inadequately accounted for or totally disregarded.

TABLE 11.3 10% Interest Factor for Annual Compounding

	Single Payment		Equal Payment Series				Uniform
	Compound-Amount	Present-Worth	Compound-Amount	Sinking-Fund	Present-Worth	Capital Recovery	Gradient-Series
	Factor	Factor	Factor	Factor	Factor	Factor	Factor
	(To find F	(To find P	(To find F	(To find A	(To find P	(To find A	(To find A
n	Given P F/P,i,n)	Given F P/F,i,n)	Given A F/A,i,n)	Given F A/F,i,n)	Given A P/A,i,n)	Given P A/G,i,n)	Given G A/G,i,n)
1	1.100	0.9091	1.000	1.0000	0.9091	1.1000	0.0000
2	1.210	0.8265	2.100	0.4762	1.7355	0.5762	0.4762
3	1.331	0.7513	3.300	0.3021	2.4869	0.4021	0.9366
4	1.464	0.6830	4.641	0.2155	3.1699	0.3155	1.3812
5	1.611	0.6209	6.105	0.1638	3.7908	0.2638	1.8101
6	1.772	0.5645	7.716	0.1296	4.3553	0.2296	2.2236
7	1.949	0.5132	9.487	0.1054	4.8684	0.2054	2.6216
8	2.114	0.4665	11.436	0.0875	5.3349	0.1875	3.0045
9	2.358	0.4241	13.579	0.0738	5.7590	0.1737	3.3724
10	2.594	0.3856	15.937	0.0628	6.1446	0.1628	3.7255
11	2.853	0.3505	18.531	0.0540	6.4951	0.1540	4.0641
12	3.138	0.3186	21.384	0.0468	6.8137	0.1468	4.3884
13	3.452	0.2897	24.523	0.0408	7.1034	0.1408	4.6988
14	3.798	0.2633	27.975	0.0358	7.3667	0.1358	4.9955
15	4.177	0.2394	31.772	0.0315	7.6061	0.1315	5.2789
16	4.595	0.2176	35.950	0.0278	7.8237	0.1278	5.5493
17	5.054	0.1979	40.545	0.0247	8.0216	0.1247	5.8071
18	5.560	0.1799	45.599	0.0219	8.2014	0.1219	6.0526
19	6.116	0.1635	51.159	0.0196	8.3649	0.1196	6.2861
20	6.728	0.1487	57.275	0.0175	8.5136	0.1175	6.5081
21	7.400	0.1351	64.003	0.0156	8.6487	0.1156	6.7189
22	8.140	0.1229	17.403	0.0140	8.7716	0.1140	6.9189
23	8.954	0.1117	79.543	0.0126	8.8832	0.1126	7.1085
24	9.850	0.1015	86.497	0.0113	8.9848	0.1113	7.2881
25	10.835	0.0923	96.347	0.0102	9.0771	0.1102	7.4580
26	11.918	0.0839	109.182	0.0092	9.1610	0.1092	7.6187

TABLE 11.3 10% Interest Factor for Annual Compounding _ Continued

	Single Payment		Equal Payment Series				Uniform
	Compound-Amount	Present-Worth	Compound-Amount	Sinking-Fund	Present-Worth	Capital Recovery	Gradient-Series
	Factor	Factor	Factor	Factor	Factor	Factor	Factor
	(To find F	(To find P	(To find F	(To find A	(To find P	(To find A	(To find A
n	Given P F/P,i,n)	Given F P/F,i,n)	Given A F/A,i,n)	Given F A/F,i,n)	Given A P/A,i,n)	Given P A/P,i,n)	Given G A/G,i,n)
27	13.110	0.0763	121.100	0.0083	9.2372	0.1083	7.7704
28	14.421	0.0694	134.210	0.0075	9.3066	0.1075	7.9137
29	15.863	0.0630	148.631	0.0067	9.3696	0.1067	8.0489
30	17.449	0.0521	164.494	0.0061	9.4269	0.1061	8.1762
31	19.194	0.0474	181.943	0.0055	9.4790	0.1055	8.2962
32	21.114	0.0431	201.138	0.0050	9.5264	0.1050	8.4091
33	23.225	0.0392	222.252	0.0045	9.5694	0.1045	8.5152
34	25.548	0.0356	245.447	0.0041	9.6086	0.1041	8.6149
35	28.102	0.0221	271.024	0.0037	9.6442	0.1037	8.7086
40	45.259	0.0137	442.593	0.0023	9.7791	0.1023	9.0962
45	72.890	0.0085	718.905	0.0014	9.6628	0.1014	9.3741
50	117.391	0.0053	1163.909	0.0009	9.9148	0.1009	9.5704
55	189.059	0.0033	1880.591	0.0005	9.9471	0.1005	9.7075
60	304.482	0.0020	3034.816	0.0003	9.9672	0.1003	9.8023
65	490.371	0.0013	4893.707	0.0002	9.9796	0.1002	9.8672
70	789.747	0.0008	7887.470	0.0001	9.9873	0.1002	9.9113
75	1271.695	0.0005	12708.954	0.0001	9.9921	0.1001	9.9410
80	2048.400	0.0003	20474.002	0.0001	9.9951	0.1001	9.9609
85	3298.969	0.0002	32979.690	0.0000	9.9970	0.1000	9.9742
90	5313.023	0.0002	53120.226	0.0000	9.9981	0.1000	9.9831
95	8556.676	0.0001	85556.760	0.0000	9.9988	0.1000	9.9889
100	13780.612	0.0001	137796.123	0.0000	9.9993	0.1000	9.9928

BIBLIOGRAPHY

Aghakouchak, A., and S. F. Stiemer. 2001. "Fatigue Reliability Assessment of Tubular Joints of Existing Offshore Structures." *Canadian Journal of Civil Engineering* 28:691–698.

Ament, P. C. H. 1998. "Corrosion Fatigue of Structural Steel in Sea Water." PhD study, Delft. Engineers.

American Concrete Institute. 1980. "Performance of Concrete in Marine Environment," Publication SP-65, Detroit.

API Recommended Practice 1632, "Cathodic Protection of Underground Petroleum Storage Tanks and Piping Systems."

API, Recommended Practice 2A-WSD (RP 2A-WSD), "Recommended Practice for Planning, Designing and Constructing Fixed Offshore Platforms—Working Stress Design, Twenty-First Edition," December 2000.

ASM Handbook, Vol. 13, "Corrosion of Titanium and Titanium Alloys," 675.

Au, S. K., and J. L. Beck. 1999. "A New Adaptive Importance Sampling Scheme for Reliability Calculation." *Structural Safety* 21:135–158.

von Baeckmann, W., W. Schwenk, and W. Prinz. 1997. *Hand Book of Cathodic Corrosion Protection*, 3rd ed. Gulf Professional Publishing, ISBN: 9780884150565

Bea, B., and M. J. K. Craig. 1993. "Developments in the Assessment and Requalification of Offshore Platforms." Offshore Technology Conference Proceedings 1993, Houston, Texas, May 3–6, 1993.

Booth, G. H., A. W. Cooper, P. M. Cooper, and D. S. Wakerley. 1967. *Br. Corros. J.* 2, no. 104.

British Standard BS4515-1. 2000. "Specification for Welding of Steel Pipelines on Land and Offshore, Part 1: Carbon and Carbon Manganese Steel Pipelines."

Bryan, W. T. 1970. *Designing Impressed Current Cathodic Protection Systems with Durco Anodes.* Dayton, OH: The Duriron Company.

Chess, P. M., and F. N. Spon, eds. 1998. *Cathodic Protection of Steel in Concrete.* New York, 187.

Coles, Michael. 2005. "Material Engineering and Coating." FMC Engineering Boot Camp Training, Houston.

Collins J.A. (1993), Failure of Materials in Mechanical Design Analysis, Prediction, Prevention. John Wiley & Son, New York pp. 16-78.

Conference on Corrosion and Infrastructure. 1995. *Practical Applications and Case Histories.* Baltimore: NACE International.

Corley, W. G. 1995. "Designing Corrosion Resistance into Reinforced Concrete." *Materials Performance* 34, no. 9 (September): 54–58.

Crane, A. P., ed. 1993. Corrosion of Reinforcement in Concrete Construction. Chichester, England: Ellis Horwood Ltd.

Darwin, David, Carl E. Locke Jr., Javier Balma, and Jason T. Kahrs. 1999. "Evaluation of Stainless-Steel Clad Reinforcing Bars." SL Report No. 99-3, University of Kansas Center for Research, Inc., 17.

DeGiorgi, V. G. n.d. "Influence of Seawater Composition on Corrosion Prevention System Parameters." Mechanics of Materials Branch, Code 6382, Naval Research Laboratory, Washington, DC.

Department of the US Army. 1985. Technical Manual TM 5-811-7, "Electrical Protection, Cathodic Protection."

Diamantidis, D., ed. 2001. *Probabilistic Assessment of Existing Structures.* Cachan Cedex, France: RILEM Publications.

Dillion, C. P. 1982. "Forms of Corrosion Recognition and Prevention: An Official NACE International Publication, Volume 1."

DNV. 1996. "Guideline for Offshore Structural Reliability Analysis. General Part and Applications to Jackets." Høvik, Norway: Det Norske Veritas.

Durr, C. L., and J. A. Beavers. 1998. "Techniques for Assessment of Soil Corrosivity." Corrosion '98, paper 667, NACE International.

Efthymiou M., J. W. van de Graaf, P. S. Tromans, and I. M. Hines. 1996. "Reliability Based Criteria for Fixed Steel Offshore Platforms." Proceeding of the 15th International Conference on Offshore Mechanic and Artic Engineering.

Escalante, E. 1989. "Concepts of Underground Corrosion." In *Effects of Soil Characteristics on Corrosion, ASTM STP 1013*, edited by V. Chaker and J. D. Palmer, Philadelphia: American Society for Testing and Materials.

Engineering Economy. n.d. ANSI Standard Z94.5. New York: American National Standards Institute.

Ersdal, G. 2002. "On Safety of Fixed Offshore Structures, Failure Paths and Barriers." Proceedings of OMAE 2002, the 21st International Conference on Offshore Mechanics and Artic Engineering, Oslo, Norway.

Eskijian, M., B. Gerwick, R. Heffron, D. Polly, and T. Spencer. International Workshop on Corrosion Control for Marine Structures and Pipelines—Harbour and Port Facilities.

Evans, U. R. 1960. *The Corrosion and Oxidation of Metals: Scientific Principles and Practical Applications.* New York: St. Martin's Press.

Freeze, R. A and J. A. Cherry. 1979. Groundwater. Englewood Cliffs, NJ Prentice-Hall.

Gamry Instruments. n.d. "Application Note, Getting Started with Electrochemical Corrosion Measurement." Assessed Monday December 30, 2013. **www.gamry.com**.

Gerhard, Ersdal. 2005. "Assessment of Existing Offshore Structures for Life Extension." PhD study, University of Stavanger Norway.

Goodall Electric. 1982. "The Custom VIP Cathodic Protection Rectifier." Technical bulletin, Fort Collins, CO: Goodall Electric, Inc., January.

Guedes Soares. C and Garbatov. Y, (1999), Reliability of maintained, corrosion protected plates subjected to non-linear corrosion and compressive loads, Unit of Marine Technology and Engineering, Technical University of Lisbon, Instituto Superior Técnico, Av. Rovisco Pais, 1096 Lisboa, Portugal

Gitman, L. J., Principles of Managerial Finance, New York, Harper Collins, 1991

Halmshaw, T. 1987. *Nondestructive Testing.* London: Edward Arnold, 108–215.

Huntington Alloys. 1979. "Resistance to Corrosion—Huntington Alloys Inc. Technical Publication No. S-37, 3rd ed.

International Standardization Organization. 2000. ISO/DIS 13822, "Bases for Design of Structures—Assessment of Existing Structures."

Kallaby, J. and O'Connor P.E. (1994). "An integrated approach for underwater survey and damage assessment of offshore Platforms", OTC 7487, Offshore Technology Conference Proceedings, Houston, May 1994

Kepler, Jennifer L., David Darwin, and Carl E. Locke Jr. 2000. "Evaluation of Corrosion Protection Methods for Reinforced Concrete Highway Structures."

Kirkley, Charles. 1982. *Oil and Gas Production Corrosion Control.* Petroleum Extension Services, University of Texas. ISBN-13: 9780886981105

Kvitrud, A., G. Ersdal, and R. L. Leonardsen. 2001. "On the Risk of Structural Failure on Norwegian Offshore Installations." Proceedings of ISOPE 2001, 11th International Offshore and Polar Engineering Conference, Stavanger, Norway.

Lemieux, E., and W. H. Hartt. 2006. "Galvanic Anode Current and Structure Current Demand Determination Methods for Offshore Structures." NACE International.

Liu, Y. 1996. "Modeling the Time-to-Corrosion Cracking of the Cover Concrete in Chloride Contaminated Reinforced Concrete Structures." Virginia Polytechnic Institute and State University.

Lye, Rolf E. n.d. *Anode Consumption on a Subsea X-Mas Tree*. Porsgrunn, Norway: Norsk Hydro Research Centre.

Madsen, H. O., and J. D. Sørensen. 1990. "Probability-Based Optimization of Fatigue Design Inspection and Maintenance." Presented at Int. Symp. on Offshore Structures, July 1990, University of Glasgow.

Mehta, P. Kumar. 1991. *Concrete in the Marine Environment*. New York: Elsevier Applied Sciences.

Meijers, S. J. H. 2003. "Computational Modeling of Chloride Ingress in Concrete." Delft University.

Melchers, R. E. 2001. "Assessment of Existing Structures—Approaches and Research Needs." *Journal of Structural Engineering* (April).

Milliams, Derek. 1993. "Corrosion Management." 12th Int. Corr. Cong. "Corrosion Control for Low Cost Reliability," September 19–24, 1993, 2420.

Moan T. Vardal O.T. (2001). Probabilistic assessment of fatigue reliability of existing offshore platforms. Proc. ICOSSAR 2001. Newport Beach, California. Morley, J. 1989. "A Review of Underground Corrosion of Steel Piling, Steel Construction."

Myers, J. R., and M. A. Aimone. 1980. *Corrosion Control for Underground Steel Pipelines: A Treatise on Cathodic Protection*. Franklin, OH: James R. Myers and Associates.

NACE Corrosion Engineer's Reference Book. 1980.

NACE International Task Group 018. 2001. *State-of-Art Survey on Corrosion of Steel Piling in Soil.* Houston: NACE International Publication.

NACE International Task Group T-6A-39. 2000. *Coating Used in Conjunction with Cathodic Protection.* Houston: NACE International Publication.

NACE RP 0169. n.d. "Standard Recommended Practice: Control of External Corrosion on Underground or Submerged Metallic Piping Systems."

NACE RP 0285. n.d. "Standard Recommended Practice: Corrosion Control of Underground Storage Tank Systems by Cathodic Protection."

NACE Standard RP0176-2003. 2003. "Recommended Practice (Latest Revision)—Corrosion Control of Steel Fixed Offshore Structures Associated with Petroleum Production."

NACE Standard RP-06-76. 1976. "Recommended Practice—Control of Corrosion on Steel, Fixed Offshore Platforms, Associated with Petroleum Production." April.

NACE Test Method TM 0497. n.d. "Measurement Techniques Related to Criteria for Cathodic Protection on Underground or Submerged Metallic Piping Systems."

National Association of Corrosion Engineers. 2003. "MR0175-2003: Metals for Sulfide Stress Cracking and Stress Corrosion Cracking Resistance in Sour Oilfield Environments."

National Association of Corrosion Engineers. 2003. "TMO284-2003: Evaluation of Pipeline and Pressure Vessel Steels to Hydrogen-Induced Cracking."

National Transportation Safety Board. 1996. Pipeline Accident Summary Report for Pipeline Rupture, Liquid Butane Release, and Fire, Lively, Texas, August 24, 1996.

National Transportation Safety Board. 2000. "Pipeline Accident Report for Natural Gas Pipeline Rupture and Fire Near Carlsbad, New Mexico, August 19, 2000."

Nilsson Electrical Laboratory. n.d. "4-Pin Soil Resistance Meter Instruction Manual Model 400."

Norsok Standard. 1997. "Common Requirements for Cathodic Protection M-503, Rev. 2," September.

Ogata, A. 1970. "Theory of Dispersion in a Granular Medium." *US Geol. Surv. Prof.*, Paper 411.1.

Parker, Marshall E. 1980. "Corrosion by Soils," *NACE Basic Corrosion Course*, Edited by A. Brasunus. Houston: National Association of Corrosion Engineers.

Parker, Marshall E., and Edward G. Peattie. 1999. "Pipeline Corrosion and Cathodic Protection."

Peabody, A. W. 1980. "Principles of Cathodic Protection," *NACE Basic Corrosion Course,* Edited by A. Brasunus. Houston: National Association of Corrosion Engineers.

Pierre. R. Roberge (1999), Handbook of Corrosion Engineering McGraw-Hill, New York USA, ISBN 07-076516-2.

Proceedings of the Royal Society, 114 (1824), pp 151-246 and 115 (1825), pp 328-346.

Revie, R. Winston, ed. 2000. *Uhlig's Corrosion Handbook*, 2nd ed. John Wily & Sons, Inc.

Romanoff, M. 1962. "Corrosion of Steel Pilings in Soil." National Bureau of Standards Monograph 58, US Department of Commerce, Washington, DC.

RP B401. 2005. "Cathodic Protection Design." Det Norske Veritas Industri Norge.

Salau, M. A., Ismail Adegbite, and E. E. Ikponmwosa. 2012. "Characteristic Strength of Concrete Column Reinforced with Bamboo Strips."

Santala, M. J. 1988. "Nigeria Environmental Data Summary—NEDS: Revision 0, MEPTC Report."

Semenski, Damir, and Hinko Wolf. 2005. "Risk Assessment of Structural Elements of the Offshore Gas and Oil Platforms." University of Zagreb, Faculty of Mechanical Engineering and Naval Ltd, 516–528.

Sheppard, Nora, ed. 1984. *Introduction to the Oil Pipeline Industry*, 3rd ed. University of Texas Petroleum Extension Service, sponsored by API.

Smith, Martin, Colin Bowley, and Lucian Williams. 2002. "In Situ Protection of Splash Zones." Winn & Coales Ltd, Paper 02214.

STI R892. n.d. "Recommended Practice for Corrosion Protection of Underground Piping Networks Associated with Liquid Storage and Dispensing Systems."

STI-R-972. n.d. "Recommended Practice for the Installation of Supplemental Anodes for STI-P3 USTs."

Strategic Highway Research Program. 1993. "Cathodic Protection of Reinforced Concrete Bridge Elements: A State-of-the-Art Report." SHRP-S-337, ELTECH Research Corporation.

Straub, Daniel. 2004. "Generic Approaches to Risk Based Inspection for Steel Structures." PhD study, Swiss Federal Institute of Technology, Zurich.

Summerville, Nicholas. 2004. *"Basic Reliability—An Introduction to Reliability Engineering."* AuthorHouse, 2004, ISBN 1418424188.

Transportation Research Board Executive Committee. 2006. "Manual on Service Life of Corrosion-Damaged Reinforced Concrete Bridge Superstructure Elements."

Uhlig, H. H., and A. Asphahani. 1979. "Corrosion Behaviour of Cobalt-Based Alloys in Aqueous Media." *Materials Performance* 18, no. 11, p. 9 (November).

UL 1746. n.d. "Standard for Safety: External Corrosion Protection Systems for Steel Underground Storage Tanks."

US Army. 2005. "Cathodic Protection." Technical manual, UFC 3-570-02A, 2005.

Verink, E. D., *Corrosion Economic Calculations, in Metals Handbook: Corrosion, Metals Park*, Ohio, ASM International, 1987, pp. 369–374.

Videla, H. A., and L. K. Herrera. 2009. "Understanding Microbial Inhibition of Corrosion: A Comprehensive Overview." *Int Biodeter Biodegrad* 63:896–900.

Wire Research Council. 1976. "Intergranular Corrosion of Chromium-Nickel Stainless Steel—Final Report," WRC Bulletin No. 138, New York.

Yong Bai, (2003) Marine structural Design, Elsevier Amsterdam. 1st Edition. ISBN: 0-08-043921-7.

Youping Liu (1998), Modeling the Time-to-Corrosion Cracking of the Cover Concrete in Chloride Contaminated Reinforced Concrete Structures, PhD dissertation submitted to Virginia State University, Department of Civil Engineering, Blacksburg Virginia.

INDEX

Printed in the United States
by Baker & Taylor Publisher Services